Smart Systems Integration and Simulation

Nicola Bombieri • Massimo Poncino
Graziano Pravadelli
Editors

# Smart Systems Integration and Simulation

Springer

*Editors*
Nicola Bombieri
Dipartimento di Informatica
Università di Verona
Verona, Italy

Massimo Poncino
Dipartimento di Automatica e Informatica
Politecnico di Torino
Torino, Italy

Graziano Pravadelli
Dipartimento di Informatica
Università di Verona
Verona, Italy

ISBN 978-3-319-27390-7 ISBN 978-3-319-27392-1 (eBook)
DOI 10.1007/978-3-319-27392-1

Library of Congress Control Number: 2015958892

Springer Cham Heidelberg New York Dordrecht London


Printed on acid-free paper

Springer International Publishing AG Switzerland is part of Springer Science+Business Media (www.springer.com)

# Contents

# Acronyms/Abbreviations

| | |
|---|---|
| 1/2/3-D | One/Two/Three-Dimensional |
| A2T | Automatic Abstraction Tool |
| ABACuS | Analog BehAvioural Conservative SystemC-AMS |
| AC | Alternative Current |
| AC | Approximate Computing |
| AC/DC | Rectifier converting alternating current (AC) to direct current (DC) |
| ADC | Analog/Digital Converter |
| ADS | Advanced Design System |
| AFE | Analog Front-End |
| ALS | Approximate Logic Synthesis |
| AMS | Analog-Mixed Signal |
| ANT | Algorithmic Noise Tolerance |
| APB | Advanced Peripheral Bus |
| API | Application Programming Interface |
| ARC | Application Resilience Characterization |
| ASIC | Application-Specific Integrated Circuit |
| AST | Abstract Syntax Tree |
| AT | Approximately Timed |
| AWGN | Additive White Gaussian Noise |
| BCD | Bipolar-CMOS-DMOS |
| BEM | Boundary Element Method |
| BSIM | Berkeley Short-channel IGFET Model |
| C | Capacitance |
| CAD | Computer Aided Design |
| CMI | Compact Model Interface |
| CMOS | Complementary Metal Oxide Semiconductor |
| CPU | Central Processing Unit |
| CTE | Coefficient of Expansion |
| CTI | Charge Transfer Interconnect |
| C–V | Capacitance–Voltage |
| DAC | Digital to Analog Converter |

| | |
|---|---|
| DBC | Direct Bonded Copper |
| DC | Direct Current |
| DC/AC | Inverter converting direct current (DC) to alternating current (AC) |
| DC/DC | Power converter converting a source of direct current (DC) from one voltage level to another |
| DCT | Discrete Cosine Transform |
| DDT | Deft Data Types |
| DFE | Digital Front-End |
| DLP | Digital Light Processing |
| DMOS | Double-Diffused Metal-Oxide Semiconductor |
| DMOSFET | Double-Diffused Metal-Oxide Semiconductor Field-Effect Transistor |
| DoF | Degrees-of-Freedom |
| dps | Degrees per Second |
| DSP | Digital Signal Processor |
| E | Capacity |
| EDA | Electronic Design Automation |
| EEG | ElectroEncephaloGraph |
| EFSM | Extended Finite State Machine |
| EKF | Extended Kalman Filter |
| ELN | Electrical Linear Network |
| EM | Electromagnetic |
| En | Enable |
| ESD | Energy Storage Device |
| ESL | Electronic System-Level |
| ESR | Equivalent Series Resistance |
| FA | Full Adder |
| FEA | Finite Element Analysis |
| FEM | Finite Element Method |
| FFT | Fast Fourier Transform |
| FIR | Finite Impulse Response |
| FU | Functional Unit |
| GPS | Global Positioning System |
| HDL | Hardware Description Language |
| HdS | Hardware-dependent Software |
| HDTLib | HDL-oriented Data Types LIBrary |
| HEV | Hybrid Electrical Vehicles |
| HFVR | High-Frequency Voltage Regulator |
| HIF | HDL Intermediate Format |
| HS | High-Side |
| HVAC | Heating, Ventilation and Air Conditioning |
| HW | Hardware |
| I | Current |
| IC | Integrated Circuit |
| IIR | Infinite Impulse Response |
| IMU | Inertial Measurement Unit |

| | |
|---|---|
| IP | Intellectual Property |
| IPEM | Integrated Power Electronics Module |
| L | Inductor |
| LASER | Light Amplification by Stimulated Emission of Radiation |
| LBS | Laser Beam Steering |
| LCoS | Liquid Crystal on Silicon |
| LDPC | Low Density Parity Check |
| LOA | Lower-part Or Adder |
| LS | Low-Side |
| LSB | Least Significant Bit |
| LSF | Linear Signal Flow |
| LT | Loosely Timed |
| mW | milli-Watt |
| MAC | Multiply And Accumulate |
| MBD | Model Based Design |
| MDSP | Main DSP |
| MED | Mean Error Distance |
| MEMS | Micro Electro-Mechanical Systems |
| MIPS | Microprocessor without Interlocked Pipeline Stages |
| MMIO | Memory Mapped Input/Output |
| MoC | Model of Computation |
| MOEMS | Micro-Opto-Electro-Mechanical System |
| MOR | Model Order Reduction |
| MOS | Metal-Oxide Semiconductor |
| MOSFET | Metal-Oxide-Semiconductor Field-Effect Transistor |
| MPPT | Maximum Power Point Transfer |
| MSB | Most Significant Bit |
| MSE | Mean Square Error |
| MW | Mega-Watt |
| OCCAM | Ordinary C++ Code for Analogue Models |
| ODC | Observability Don't Care |
| OPAMP | Operational Amplifier |
| OSTC | Open Source Test Case |
| PBL | Probabilistic Boolean Logic |
| PC | Probabilistic Computing |
| PCB | Printed Circuit Board |
| Pcell | Parameterized layout Cell |
| PCMOS | Probabilistic CMOS |
| PDK | Process Design Kits |
| QCC | Quality Constraint Circuit |
| QPSK | Quadrature Phase Shift Keying |
| R | Resistor |
| RC | Resistor Capacitor |
| RF | Radio Frequency |
| RFID | Radio Frequency Identification |

| | |
|---|---|
| RLC | A resistor (R), an inductor (L), and a capacitor (C) connected in series or in parallel |
| ROM | Reduced-Order-Model |
| RPR | Reduced Precision Redundancy |
| RTL | Register Transfer Level |
| SALSA | Systematic methodology for Automatic Logic Synthesis of Approximate circuits |
| SiP | System-in-Package |
| SC | Stochastic Computing |
| SCNSL | SystemC Network Simulation Library |
| SDD | Significance Driven Design |
| SDE | Sentaurus Device Editor |
| SNR | Signal-to-Noise Ratio |
| SoB | System on Board |
| SOC | State Of Charge |
| SOI | Silicon on insulator |
| SPICE | Simulation Program with Integrated Circuit Emphasis |
| SW | Software |
| SVD | Singular Value Decomposition |
| TCAD | Technology Computer Aided Design |
| TCP/IP | Transmission Control Protocol/Internet Protocol |
| TDF | Timed Data-Flow |
| TIA | Transimpedance Amplifier |
| TLM | Transaction Level Modeling |
| TPMS | Tire Pressure Monitoring System |
| TSV | Through-Silicon Vias |
| UART | Universal Asynchronous Receiver-Transmitter |
| UAV | Unmanned Aerial Vehicle |
| UIS | Unclamped Inductive Switching |
| UNIVERCM | UNIversal VERsatile Computational Model |
| UVM | Universal Verification Methodology |
| UWB | Ultra-Wide Band |
| V | Voltage |
| VHDL | VHSIC Hardware Description Language |
| VHSIC | Very High Speed Integrated Circuit |
| VOS | Voltage Over-Scaling |
| μW | micro-Watt |

# Chapter 1
# Introduction

**Nicola Bombieri and Graziano Pravadelli**

Smart Systems represent a broad class of systems defined as intelligent, miniaturized devices incorporating functionality like sensing, actuation, and control. In order to support these functions, they must include sophisticated and heterogeneous components and subsystems such as application-specific sensors and actuators, multiple power sources and storage devices, intelligence in the form of power management, baseband computation, digital signal processing, power actuators, and subsystems for various types of wireless connectivity.

It is evident from this heterogeneity that Smart Systems leverage a variety of different technologies and different materials. The challenge in the implementation of Smart Systems goes therefore beyond the design of their individual components (an already difficult task by itself), and rather lies in the co-existence of a multitude of functionalities, technologies, and materials. The widely acknowledged keyword in Smart Systems design is in fact integration. There are essentially two dimensions of integration that represent the main obstacle towards mainstream design of Smart Systems: Technological and methodological. As already experienced in specific domains (e.g., in digital and analog design), a solution has been found first for the technological issues. Advanced packaging technologies such as System-in-Package (SiP) and chip stacking (3D IC) with through-silicon vias (TSVs) allow today manufacturers to package all this functionality more densely, combining various technological domains in a single package. SiP technology works nicely because it allows merging components and subsystems with different processes, and mixed technologies using the state-of-the-art advanced IC packaging technologies with minor impact on the design flow. Therefore, to some extent, technological solutions aimed towards integration are available.

---

N. Bombieri (✉) • G. Pravadelli
University of Verona, Verona, Italy
e-mail: nicola.bombieri@univr.it; graziano.pravadelli@univr.it

 
N. Bombieri et al. (eds.), *Smart Systems Integration and Simulation*,
DOI 10.1007/978-3-319-27392-1_1

Design methodologies, however, are falling behind: Current design approaches for Smart Systems use separate design tools and ad-hoc methods for transferring the non-digital domain to that of IC design and verification tools, which are more consolidated and fully automated. This solution is clearly sub-optimal and cannot respond to challenges such as time-to-market and request of advanced sensing functionalities. A big step towards effective large-scale design of Smart Systems would be that of changing their design process from an expert methodology to a mainstream (i.e., automated, integrated, reliable, and repeatable) design methodology, so that design costs are reduced, time-to-market is shortened, design of the various domains is no longer confined to teams of specialists inside IDMs and system miniaturization can be achieved with limited risks. This objective can be reached by defining and implementing a structured design approach that explicitly accounts for integration as a specific constraint.

This book aims at achieving this ambitious objective by proposing a holistic and multidisciplinary co-design approach, which requires closing several technical and cultural gaps. The proposed solution is the result of a joint co-operation of research and industry partners, including EDA vendors, to ensure the approach applicability in realistic, industry-strength design flows and environments, with a direct impact on the industrial exploitation.

Behind the growing interest in Smart Systems, there is a potentially huge and quickly growing market, which is expected to grow in the order of $200B in 2020, inducing an even larger market of non-hardware services involving all the various devices envisioned in the Internet of Things. Such a market is much larger than those of smart- or feature-phones in terms of number of devices. Over 50 billion devices will be connected to the Internet according to Cisco forecasts, and most of these devices will be Smart Systems. Miniaturized Smart Systems find applications in a broader range of key strategic sectors, including automotive, healthcare, ICT, safety and security, and aerospace. Also, efficient energy management and environment protection are business sectors in which the utilization of miniaturized Smart Systems may make a difference. The worldwide market for "Monitoring & Control" products and solutions, one of the most important fields of Smart Systems applications, containing solutions for environment, critical infrastructures, manufacturing and process industry, buildings and homes, household appliances, vehicles, logistics and transport or power grids, is around 188B Euro. This value represents 8 % of the total ICT expenditures worldwide, and it is identical to the whole semiconductor industry world revenues and approximately twice that of the world mobile phone manufacturers revenues.

In Chap. 2, the book introduces the general definition of a smart electronic system, providing a list of real-world examples to show today's application scenarios. The chapter analyses the complexity of Smart Systems, describing the technology domains of the typical components and the main system architectures. The chapter concludes with a description of the main challenges for Smart Systems designers, which essentially involves the integration of a number of functionalities, materials, and technologies in a single device.

Chapter 3 deals with simulation domains and abstraction levels in Smart Systems design. The high level of heterogeneity of Smart Systems makes design a very challenging task, as each domain is supported by specific languages, modelling formalisms, and simulation frameworks. A major issue is furthermore posed by simulation, that heavily impacts the design and verification loop and that is further complicated by such a heterogeneous context. Chapter 3 provides a formalization of the typical abstraction levels and design domains of a Smart System. This taxonomy allows identifying a precise role in the design flow for co-simulation and simulation scenarios, and to examine the impact of heterogeneous or homogeneous models of computation. It proposes a methodology to move from the co-simulated heterogeneity to a simulatable homogeneous representation of the entire Smart System in C++ (or its extension SystemC-AMS). The code generation methodology supports both digital and circuit-level descriptions, and it is enhanced by scheduling optimizations to improve the effectiveness of the generated code.

Smart Systems applications often include error resilient computations, due to the presence of noisy input data, the lack of a unique golden output, etc. Therefore, computation accuracy constraints can be relaxed to improve a system efficiency. Recently, a design paradigm called Approximate Computing (AC) has been proposed to formalize the exploitation of the accuracy dimension as a way to optimize efficiency in digital computing systems. AC configures as one of the most promising ways to reduce energy consumption in Smart Systems. Chapter 4 presents an overview of the different AC techniques proposed in literature. The chapter focuses on Algorithmic Noise Tolerance (ANT), one of the most suitable AC approaches for Smart Systems applications. In particular, it presents an analysis of the automatic application of this technique to an existing design. It shows how this automation can be achieved with a flow that leverages standard EDA tools, with minimal input from the designer. It concludes by showing how the proposed technique allows obtaining almost 45 % total power saving for a typical DSP circuit.

Chapter 5 purpose is to present a flexible software platform for design, integration, and co-simulation of power components by enabling multi-domain simulations for a variety of power devices and modules. A few of models are built for power devices, and the integration between multiple physical domains, like the electrical and the thermal ones, is obtained. In particular three main examples are developed: (1) the definition of a complete methodology to implement a physics-based macro-model of a power module (Diodes, IGBTs, or MOSFETs) able to perform the co-simulation of the thermal and electrical phenomena; (2) the exploitation of an innovative automatic synthesis flow aimed at extracting compact electro-thermal models for power MOSFETs; (3) the FEM analysis of power MOSFET failures during UIS (unclamped inductive switching).

Chapter 6 deals with system-level modelling of Micro-Electro-Mechanical-Systems (MEMS). It presents an integrated approach of simulating and optimizing MEMS based systems while taking into account the multi-physical behaviour between different domains, such as the MEMS device itself, the electronic readout circuitry, the packaging or the control software. The chapter describes a design methodology and its software implementation that overcomes the limitations inherent to methods that concentrate only on one of these domains. Parameterized and interoperable MEMS models can be generated via a library approach. Subsequent model-order-reduction techniques are applied to automate the export of MEMS models in standard hardware description languages in order to reduce the complexity and decrease simulation time by orders of magnitude allowing rapid system design.

Chapter 7 covers the energy storage and power sources domain. Power sources are an essential component in applications involving portable and wearable electronics, which are intended to be self-powered for possibly long periods of time. These devices are indeed assumed to be energy-autonomous, and besides minimizing their power consumption, they should also make careful use of the energy scavenged from the environment, namely how the latter is converted, stored, and dispatched to the components that consume it. The chapter highlights the main characteristics and challenges in the design of the power dimension of a Smart System. The chapter details models and techniques suitable for validating power sources via simulation, thus enhancing design space exploration and system dimensioning. Finally, the chapter proposes a semi-automatic methodology for efficient power source simulation based on the standard language SystemC-AMS, that allows easy integration with functional Smart System design, still guaranteeing high accuracy and efficient simulation.

Chapter 8 presents two case studies showing how the proposed approach applies to Smart Systems design and optimization. The former is the virtual prototyping platform built for a laser pico-projector actuator, where MEMS, analog, and digital components are simulated with the aim of optimizing the resulting image quality by means of firmware tuning. The latter, in the context of wearable equipment for inertial body motion reconstruction, deals with the modelling of an inertial sensor node, supporting system accuracy evaluation and sensor fusion enhancement.

By covering theoretical and practical aspects of Smart Systems design, the book targets people who are working and studying on hardware/software modelling, component integration and simulation under different positions (system integrators, designers, developers, researchers, teachers, students, etc.). In particular, it is a good introduction to people who have interest in managing heterogeneous components in an efficient and effective way on different domains and different abstraction levels. People active in Smart Systems development can understand both the current status of practice and future research directions.

# Chapter 2
# Smart Electronic Systems: An Overview

**Alessandro Sassone, Michelangelo Grosso, Massimo Poncino, and Enrico Macii**

## 2.1 Introduction

The term *smart systems* is quite general and identifies a broad class of intelligent and miniaturized devices that are usually energy-autonomous and ubiquitously connected. They incorporate functionalities like sensing, actuation, and control. In order to support these functions, they must include sophisticated and heterogeneous components and subsystems, such as digital signal processing devices, analog devices for RF and wireless communication, discrete elements, application-specific sensors and actuators, energy sources, and energy storage devices (as shown in Fig. 2.1).

These systems take advantage of the progress achieved in miniaturization of electronic systems, and are highly energy-efficient and increasingly often energy-autonomous, and can communicate with their environment.

Thanks to their heterogeneous nature, smart embedded and cyber-physical applications are able to deliver a wide range of services, and their application may lead to provide solutions to address the grand social, economic, and environmental challenges such as environmental and pollution control, energy efficiency at various scales, aging populations and demographic change, risk of industrial decline, security from micro- to macro-level, safety in transportation, increased needs for the mobility of people and goods, health and lifestyle improvements, just to name the most relevant [19].

A. Sassone • M. Poncino (✉) • E. Macii
Dipartimento di Automatica e Informatica (DAUIN), Politecnico di Torino, C.so. Duca Degli Abruzzi 24, I-10129 Torino, Italy
e-mail: alessandro.sassone@polito.it; massimo.poncino@polito.it; enrico.macii@polito.it

M. Grosso
ST-Polito s.c.ar.l., C.so. Duca Degli Abruzzi 24, I-10129 Torino, Italy
e-mail: michelangelo.grosso@st-polito.com



N. Bombieri et al. (eds.), *Smart Systems Integration and Simulation*,
DOI 10.1007/978-3-319-27392-1_2

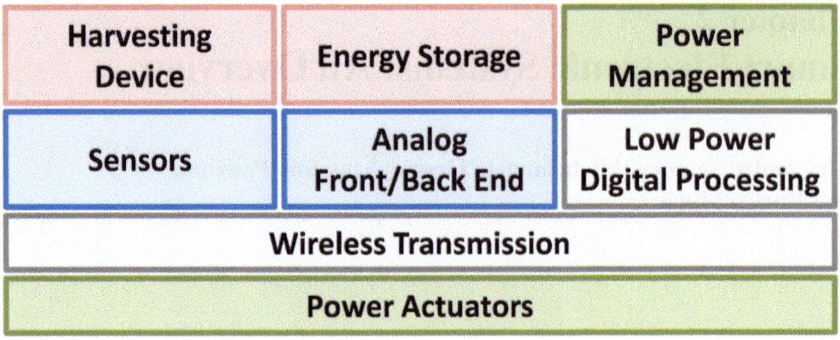


**Fig. 2.1** Typical components and domains of a smart electronic system

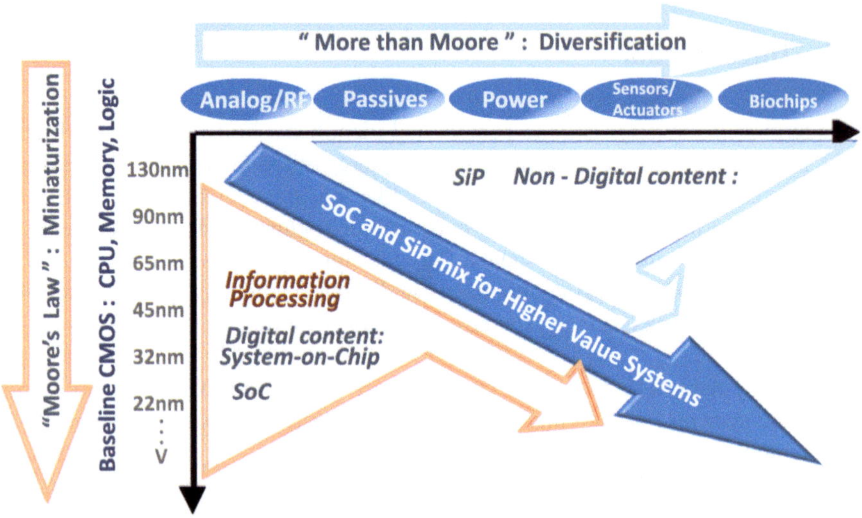


**Fig. 2.2** Smart systems as the convergence of more Moore and more than Moore

The deployment of smart systems requires the concurrence of technological advancements both in the dimension of (1) increased diversification and heterogeneity in a single device, i.e., technologies such as 3D integration, Systems-in-Package, and (2) increased integration capabilities of silicon technologies. These two dimensions are the well-known technology paths described in the International Technology Roadmap for Semiconductors (ITRS) and that are traditionally considered as complementary directions. In that respect, smart systems as a category of devices represent the convergence between these two dimensions, as depicted in Fig. 2.2 [25].

Recent years have witnessed a significant increase of the relative weight of the "More-than-Moore" component in the industry evolution; this will allow to establish a new "virtuous cycle" no longer based solely on device scaling but relying on other innovations at the system, technology, circuit, and device levels, which will need to address not just front-end technologies but also backend/packaging ones. Such opportunities are counterbalanced by challenges related to the integration and manufacturing of such devices, the development of tools and flows for the automated design of these systems, as well as the coordination of design teams that reflect the heterogeneity of smart system in their competences (mechanics, electronics, communication, optics, computer science, etc.).

### *2.1.1 Evolution of Smart Systems*

The smart systems community traditionally classifies smart systems by their degree of autonomy, meant as both in terms of reduced need of external supervision and control and of energy self-sufficiency. This has brought to the traditional three-level categorization into "generations" of smart systems [19], corresponding to different extents to which the key functionalities are implemented.

**First Generation Smart Systems** integrate sensing and or actuation as well as signal processing to enable various types of actions. Such systems have already been successfully deployed in many application domains, e.g., personal devices to monitor the health status of persons or safety systems in automotive applications.

**Second Generation Smart Systems** add perception features and are predictive and adaptive systems, possibly with self-test capabilities. Moreover, they include network connectivity of some type and advanced energy scavenging and management capabilities. An example of this category are smart RFID labels that are able to measure multiple parameters (temperature, acceleration, etc.) for transport real-time monitoring.

**Third Generation Smart Systems** add human-like perception on top of second generation systems. They interact independently and without external control or decision, and implement systematically features like self-calibration, self-test, and self-healing. An autonomously driving car is the most typical example of such a system.

Systems of different generation evolve at different speeds; while for first-generation systems their optimization is still an R&D challenge, in second- and third-generation systems still basic scientific, materials and manufacturing challenges have to be assessed. The co-existence of different generations of systems suggests that adopting a comprehensive design approach will be likely to yield competitive products today and keep them competitive in the future versions.

This chapter presents an overview of three relevant aspects of smart systems: the application domains (Sect. 2.2), their architectures (Sect. 2.3), and the main design challenges (Sect. 2.4), from a system-level perspective and with the objective of providing a sort of guided path through the topics dealt in the remainder of the book.

## 2.2 Application Domains

The current technology and market trends show that smart systems are used in an increasingly wide range of contexts and environments, from everyday life tasks to highly complex and critical missions. The architecture and the implementation of each system needs to comply with the application requirements and constraints in terms of functionality, performance, dependability, autonomy, and safety, while keeping design and manufacturing costs as low as possible.

In the following, a list of application domain categories is reported. The purpose of this list is the identification of the smart system application domains of today and the immediate future, with the aim of being as much general and exhaustive as possible. Actual examples of smart systems are provided for each category: some of them are concepts only, others are prototype boards or products, and others have already reached the status of miniaturized devices.

### *2.2.1 Transportation*

Transportation scenarios include land (i.e., road, off-road, rail, cable, and pipeline), sea, and air (i.e., air and space) for humans and goods in general. Smart systems can be used within a moving vehicle or can take part of infrastructures and networks for transportation enhancement and regulation. Some smart device examples in this category include heterogeneous sensor-based GPS enhancement systems, engine sensing and control systems, electronic stability control systems for vehicles, parking sensors [24], car theft detection and monitoring devices, airplane balancing aids, vibration analysis instrumentation for model optimization and unmanned aerial vehicle (UAV) attitude, and operating controls.

Figure 2.3a shows a smart system example in this application domain, i.e., an accelerometer-enhanced tire pressure monitoring system (TPMS) [8, 20]). Such system measures important dynamic variables, e.g., forces, load transfer, actual tire road friction (kinetic friction), and maximum tire road friction available (potential friction), in order to actively guarantee the car safety. The data acquisition system is based on a distributed architecture composed of a number of complex intelligent sensors inside the tire that form a wireless sensor network with coordination nodes placed on the body of the car. The single sensor node represented in Fig. 2.3b is composed of MEMS sensors, analog and digital circuits (including a microcontroller), and a UWB radio link. In addition, the sensor node energy is supplied by means of harvesting, which may be based on electromechanical vibrational energy or on electromagnetic coupling with an external illuminator.

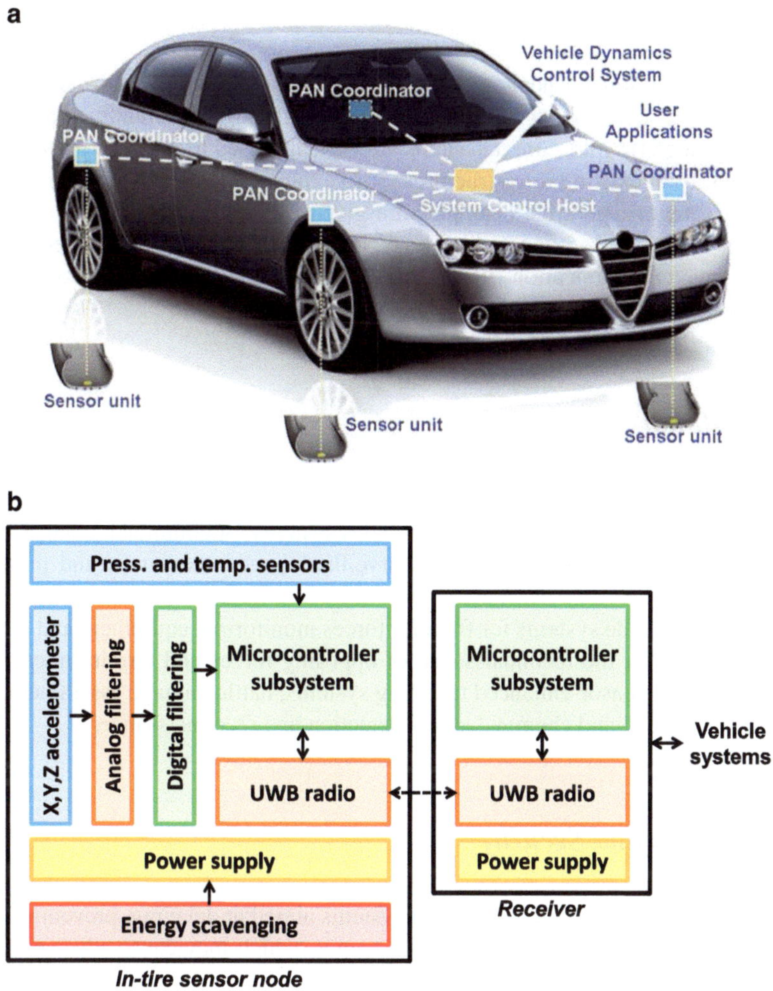


**Fig. 2.3** Intelligent tire system presented in [8]: (**a**) overview of the system architecture and (**b**) architecture of the sensor node-receiver system

The design of this system and especially of the in-tire sensor node is extremely challenging due to the very limited available energy combined with strict application requirements for data rate, delay, size, weight, and reliability in a highly dynamic environment. The integration of the sensor node is especially critical due to the harsh environment, low-power constraints, and limited size: a compact circuit board hosts all the sensor node components.

### 2.2.2 Telecommunications

This application domain deals with the generic transmission of information (e.g., audio, video, text, data from sensors, alarm events) through different media, such as cable, air, water, or void. Smart systems are typically used to improve performance and reliability of existing infrastructures, or to explore new areas and potentials.

In this category, possible examples are given by MEMS fiber-optics switches, energy harvesting and autonomous radio repeaters [5], optical MEMS-based devices for lightwave communication [9], and RF MEMS-based tunable filters and antennas [23].

### 2.2.3 Military and Defense

This application field includes devices used in defending countries from threats both foreign and domestic and specifically includes systems for command, control, communications, computing, intelligence, surveillance, reconnaissance, and targeting (C4ISRT).

Examples include systems for friendly forces monitoring, equipment and ammunition, systems for the reconnaissance of opposing forces and terrain, microrobot surveillance [3], sensor-enhanced targeting systems, battle damage assessment, and nuclear, biological, and chemical attack detection and reconnaissance.

### 2.2.4 Safety and Security

This application domain includes smart systems aimed at delaying, preventing, and otherwise protecting against accidents or crimes, which may cause adverse effects to people or organizations.

Examples include free fall sensors and human airbags [17], chemicals/radiation sensing systems, and anti-theft and anti-intrusion sensor systems.

Figure 2.4a shows an example of mobile human airbag, i.e., a protective system based on the idea of automobile airbags systems [17]. This system has two modules: a sensing module ($\mu$IMU) and an inflator that connects to two nylon airbags. When the $\mu$IMU detects a fall, it triggers an inflator, which then deploys the airbags before impact to protect the human body. The gas is supplied from a handy compressed gas cylinder, rather than the combustion of chemicals. The main components of the independent mobile airbag system are a set of MEMS sensors (accelerometers and gyroscopes), an embedded microcontroller unit performing DSP functions, and a mechanical airbag deployment system. The system is powered by means of lithium batteries. The sensors are directly connected to the A/D converters inside the microcontroller. The mechanical part includes the inflator structure for compression,

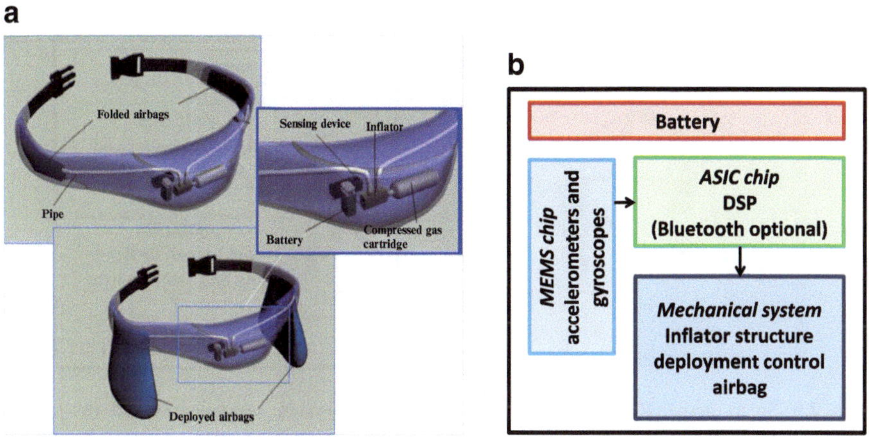


**Fig. 2.4** Intelligent mobile human airbag system of Shi et al. [17]: (**a**) system components and (**b**) system architecture

airbag deployment control, and airbag design. Figure 2.4b provides a high-level architectural view of the system.

Design and simulation of this type of system require merging non-miniaturized mechanical parts and electronic components (sensors and digital hardware including the microcontroller and the software), also taking into account the real-time application requirements.

### *2.2.5 Home Automation*

Smart systems aimed at improving convenience, comfort, energy efficiency, and security to residential buildings fall into this category, also known as domotics.

Examples are energy-efficient distributed heating, ventilation and air conditioning (HVAC) sensing and control systems like the energy management system presented in [14], acoustic monitoring systems, audio/visual switching and distribution systems, and light control systems.

### *2.2.6 Industrial Automation and Logistics*

Industrial automation deals with the optimization of energy-efficient manufacturing systems by precise measurement and control technologies. Logistics concentrates on the flow of goods between the point of origin and the point of destination to meet the requirements of customers and corporations, and it involves the integration

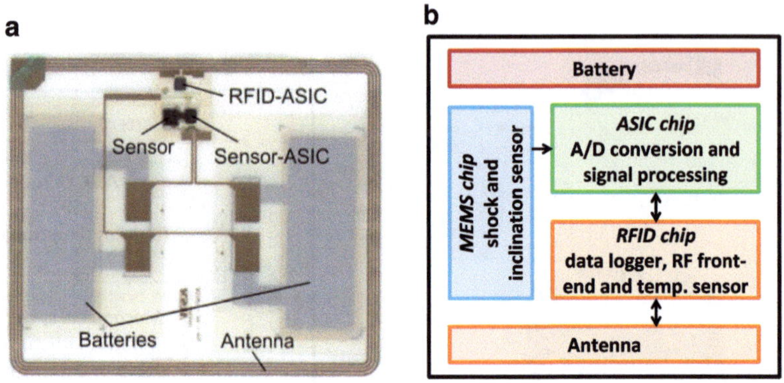


**Fig. 2.5** Smart active RFID label presented in [8]: (**a**) system components and (**b**) system architecture

(and interaction) of information, transportation, inventory, warehousing, material handling, and packaging, and often security.

This category of smart systems includes examples such as sensor-enhanced robot controls, manufacturing plant monitoring systems, and active RFID tags as the one showed in Fig. 2.5a [16].

Such active radio frequency identification label is designed for the monitoring of shock, inclination, and temperature during transportation processes. The system architecture is shown in Fig. 2.5b. Besides the components of a passive label "RF front-end, memory, and antenna" the system contains a MEMS inertial sensor system, a temperature sensor, a data logger, and a printed battery. The battery works as autonomous energy source for the inertial sensor system, the temperature sensor, and the data logger. The energy for the radio communication of the label is generated by the antenna from the electric field emitted by the reader.

All these elements have to be integrated on the thin label substrate fulfilling specific requirements for the sensor system like low power consumption, high signal to noise ratio, high temperature stability, and low device/sensor thickness.

### 2.2.7 *Laboratory Equipment*

Today's integrated advanced technologies allow the fabrication of compact, accurate, and energy-efficient instrumentation to be used for analysis, measurement, and manipulation in a wide range of fields.

In this category, possible examples are spectrometers and interferometers [16], MEMS scanners and projectors [9], and intelligent motion surfaces for manipulation such as [26].

### 2.2.8 Environment and Food/Beverage

Multi-sensing microsystems can be used for environmental applications (including monitoring and treatment) or food and beverage quality and safety.

Examples in this category include sensor nodes (within networks) for environment monitoring (e.g., to study influences on crops, livestock) based on system-on-chip design [2]. Other examples are also general-purpose microsensor modules [7], macroinstruments for large-scale earth monitoring and planetary exploration, forest fire detection and flood detection sensors, animal movement tracking systems, and tagging/tracking in supply chains (RFID).

### 2.2.9 Healthcare and Biomedical

This category includes systems aimed at generally improving health, by means of delivering diagnosis, treatment, care, and support of patients in healthcare systems. Typically different disciplines are involved in this field, including biology, genetics, physiology, physics, and bioengineering. Given the increasing importance of these application domains, various examples are available, such as biochips, microinstrumentation for microinjection and cell-manipulation, microsystems interacting with the human body, lab on chips including sensors and active microfluidics, systems for tele-monitoring of human physiological data, tracking and monitoring systems for doctors and patients inside a hospital, drug administration systems, and smart textiles for health monitoring.

Among the available case studies, it is worth mentioning the self-powered wireless pulse oximeter presented in [21], a wearable battery-free wireless electroencephalograph (EEG) [22], breath monitoring systems, limb tracking systems, wireless multi-sensor microsystems for human physiological data monitoring [4], and wearable posture corrective systems using biofeedback [10].

### 2.2.10 Power Generation, Distribution, and Harvesting

In this application scenario smart systems convert energy from different sources, store and distribute electricity to the users. Energy harvesting usually refers to the process by which energy is derived from external sources, captured and stored for small, autonomous devices at a low scale.

Sensor-based systems for control of wind turbines and portable, multipurpose energy harvesting bracelets [18] are just some examples of this increasingly studied field in the today's efforts aimed at lowering the carbon footprint.

### 2.2.11 Consumer Applications

Consumer electronics refer to equipment intended for everyday use, most often in entertainment, communications, and office productivity. This market segment includes cellular communication, personal computing, digital photography, multimedia and entertainment, fitness appliances and gaming, and it is characterized by fast cycle time, low product price and wide diffusion.

Smart system examples in this category include wireless sensor devices for statistical data logging (e.g., on balls, players, field, etc.), general-purpose microsensor modules [7], impact-sensing accelerometer systems for sport helmets [13], sailing sports wind analysis systems, vibration reduction systems in sporting goods (tennis rackets, golf clubs, etc.), smart lighting modules, augmented reality devices [12], interactive screens and interactive museums.

New smart lighting modules are becoming popular in lighting appliances for building environments.The smart lighting modules are multi featured devices that extend their functions beyond the lighting, by adding security, safety, comfort, and wireless control functions to the lighting devices. To provide this broad range of functionalities, smart lighting modules integrate components from different technological domains, hence, they can be included in the class of smart electronic systems. A real example is the smart bulb showed in Fig. 2.6.

The smart bulb is a fully controllable light bulb, embedding light changing features, complex RF communication, sensors and actuators. The bulb's light source, heat sink, and lens system are integrated with the sensors and actuators. An electronic controllable light shape system is also typically integrated in the bulb. The smart bulb has wireless secure communication channels and is remotely

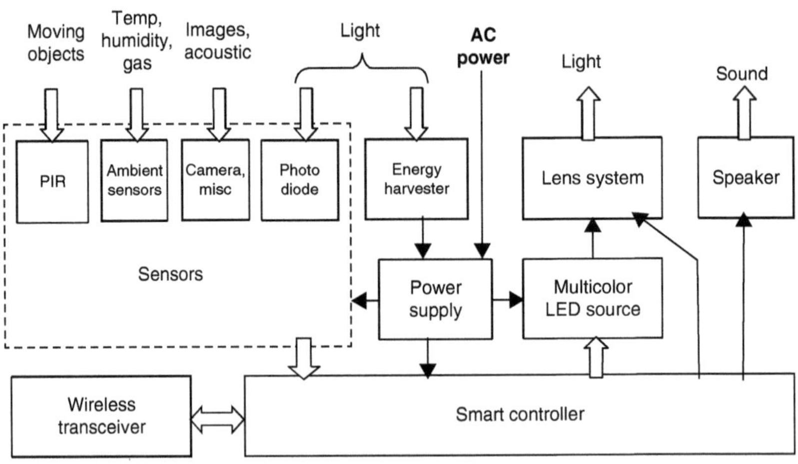


**Fig. 2.6** The smart bulb architecture

controllable, also acting as a network device capable of routing data packets and related information to other network devices. The multi featured bulb opens new functions in the lighting sector expanding it beyond the state of the art, which is up to now represented by the Philips lighting "HUE" bulb [15].

## 2.3 Smart System Architectures

### *2.3.1 Module-Level*

As mentioned above, smart embedded systems incorporate heterogeneous components, i.e., from different technology domains and providing various functions. For all systems the basic building blocks are conceptually similar, however, in each instance the specific implementation can greatly differ. A general classification of the basic building blocks is the following (see Fig. 2.7):

- *Energy Source*: Harvesting devices capable of converting energy of a physical source into electricity, such as solar (e.g., photovoltaic cells), thermal (e.g., thermoelectric energy generators), and mechanical (e.g., piezoelectric scavenger) energy generators.
- *Energy Storage*: Devices capable of storing a limited amount of electrical energy in the potential, kinetic, chemical, or other forms of energy, and restoring the stored energy back to the electrical energy on demand. The main types of energy storage devices which are generally used for smart embedded systems are batteries, supercapacitors (or ultracapacitors), and fuel cells.
- *Energy Conversion*: Components that in general convert electric energy from one form to another. Their functionality is fundamental in order to transfer the energy within the system and hence to realize the power supply to all components of the system. Energy conversion devices can be typically divided into DC–DC, AC–DC, DC–AC, and AC–AC converters.
- *Power Devices*: Energy management components such as power diodes, thyristors, power FETs, and power MOSFETs.

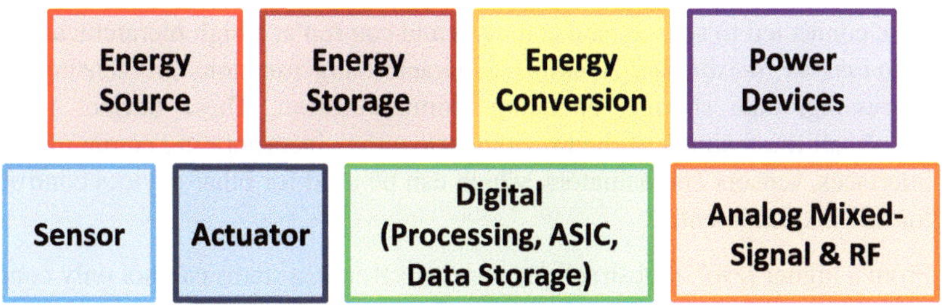


**Fig. 2.7** Basic components included in a smart system

- *Sensor*: Devices capable of detecting events or changes of a physical quantity and converting them into an electrical signal. Examples are MEMS, electro-optical sensors, image sensors, thermocouples, and acoustic sensors.
- *Actuator*: Devices capable of converting an electrical signal into another form of energy, such as electric motors, light-emitting diodes, and loudspeakers.
- *Digital*: Digital hardware blocks for processing and storing digital information, such as processor or digital signal processing (DSP) cores, digital accelerators, device controllers, and also application-specific ASICs. This category includes also the embedded software executed by the hardware blocks.
- *Analog Mixed-Signal and RF*: Analog components such as RF communication devices, signal conditioning, and interface circuits.

### 2.3.2 System-Level

Given the general classification of basic components defined above, smart systems can also be categorized based on their general module architecture and main features. Each miniaturized intelligent system falls in one of the following categories, defined by a set of characterizing functions.

- *Sensor Node (Within a Network)*: Characterizing functions are sensing, data processing, data storage, communication. A sensor node is a device that acquires data from the environment, optionally performs some kind of elaboration, and either stores data or directly transmits it to other devices (usually through a wireless channel). Low power consumption and maintainability are very common requirements, especially for remote devices.
- *Actuator Node*: Characterizing functions are data processing, actuating, communication. This type of device runs operations (e.g., turning on or off some other device) when programmed or when it receives the required command from the network. Reliability is fundamental when safety- or mission-critical tasks are executed.
- *Communication Node*: Characterizing functions are data processing, data storage, communication. This device communicates within a network and optionally elaborates data. It can operate as a remote database, a repeater, or an intelligent node connected to sensors and actuators and can run at a high hierarchical level.
- *Autonomous Sensor and Actuator*: Characterizing functions are sensing, data processing, data storage, actuating, communication. These devices include mainly all the features of the previous categories, incorporating communication interfaces, sensors and actuators, which can be used for other devices control or for self-displacement.

From a higher level of abstraction, smart electronic systems can not only consist of a single heterogeneous device, but can also be arranged in various architectures with different degrees of complexity. The following categories summarize the main system-level architectures characterizing the example devices showed in the previous section.

- *Single Module*: A single-module smart system can perform all operations related to its purpose without communicating to a host or other devices.
- *Host-Client System*: A host-client system is based on two smart (sub-) system modules, where one typically is used to access the information stored or elaborated by the other, or to program or control it.
- *Network*: The system comprises many devices, either communicating among them or connected through a network that may also be built on a hierarchical model. The devices can share a common module architecture or can be heterogeneous (e.g., many sensor nodes and an intelligent data collector node). Sensor networks, which represent one of the most common application of smart systems, are surveyed in [1] and [11].

## 2.4 Design Challenges

Smart embedded systems are produced with very different technologies and materials. This is even more evident from the heterogeneous categories of components listed in Sect. 2.3.1. Therefore, besides the design of the individual components and subsystems, the main challenge for smart system designers lies in the integration of a number of functionalities, materials, and technologies. In general, there are essentially two dimensions of integration that represent the main obstacle towards mainstream design of smart systems: *technological* and *methodological*.

As already experienced in other domains (e.g., digital and analog design), technological issues have been the first concern for research and industrial actors. Today manufacturers are able to package all aforementioned components more densely, combining the various domains in a single package. This is possible thanks to advanced packaging technologies such as System-in-Package (SiP) and chip stacking (3D IC) with through-silicon vias (TSVs). SiP technology is a good solution because it allows combining components and subsystems with different processes, and mixed technologies using the state-of-the-art advanced IC packaging technologies with minor impact on the IC chip design flow.

Nevertheless, the assembly of intrinsically heterogeneous components and the continuous miniaturization and integration push, motivated by the increasing market demand for faster, cheaper, and more performing devices, raise the need for new design and simulation methodologies. Such methodologies are fundamental for exploring the design space in order to find the most efficient trade-off between performance and involved resources, and for evaluating and validating system behavior taking into account the interactions between closely coupled components of different nature.

Considering the development of a complex system it is possible to identify several main steps characterized by their own specific peculiarities: the architecture definition, the design implementation and its validation, and the product engineering and industrialization.

During the design phase the application specifications are analyzed, the appropriate system architecture is defined usually following a top–down approach and the appropriate building blocks are identified when already available, or otherwise requested to be developed. The development team is asked to propose a solution able to cover the application functionalities supported by a detailed feasibility study. To achieve this task in a sustainable way, similar systems are evaluated when possible, expected system performances are addressed, subsystems and building components are detailed and final system cost forecasted. The mentioned feasibility analysis is of primary importance because it is the pillar on which company management takes the decision to proceed or not in the development. Considering the rhythm on which new high-tech complex products are presented on the market, to grant an appropriate level of competitiveness another key factor is the ability to have fast and effective evaluation about the new addressed solution especially on its innovative features. It is now clear the need of a platform able to manage the smart system complexity and provide quick but pertinent feedbacks about functionalities, fair heterogeneous component interoperability, reasonable forecast on sensible parameter working ranges (i.e., speed, power consumption, autonomy, efficiency, etc.), adherence on imposed regulations or applicable standards, and at least a rough estimation of underlying costs. It cannot be neglected the fact that in the feasibility phase several different proposals, architectures, and technologies are often addressed, many metrics and complex figures of merit are adopted to identify, from the beginning, the most promising approach. Also for this purpose, the capability to rapidly evaluate different scenarios, replacing building blocks, trying different basic components represent an obvious advantage in order to achieve a better awareness for all the further development steps.

System integrators typically have separate tools to model the environment. Design requires merging heterogeneous units from cross-sectional, separate, and so far loosely correlated domains. Subsystems are designed based on diverse assumptions and techniques, and are typically modeled with digital, multi-physics, or analog models, which are available at specialized design houses and silicon makers in various forms. The involved components are usually described using different languages, relying on different models of computation and modeling parameters, and need to be jointly simulated at various abstraction levels.

Figure 2.8 shows an example of operating scenario for a typical smart embedded system. The functional interactions within the components and between the components and the environment must be considered when designing the system: among them there are analog signals, digital interfaces, mechanical interactions (e.g., a sensor transfer function). In addition to this it is increasingly important to consider non-functional interactions that may have an impact on the behavior of the complete system, either on its specific functionalities or on power consumption, life duration, etc.: they may correspond to mechanical interference, heat flow, RF interference, etc. In Fig. 2.8, only thermal flows have been reported as non-functional interactions,

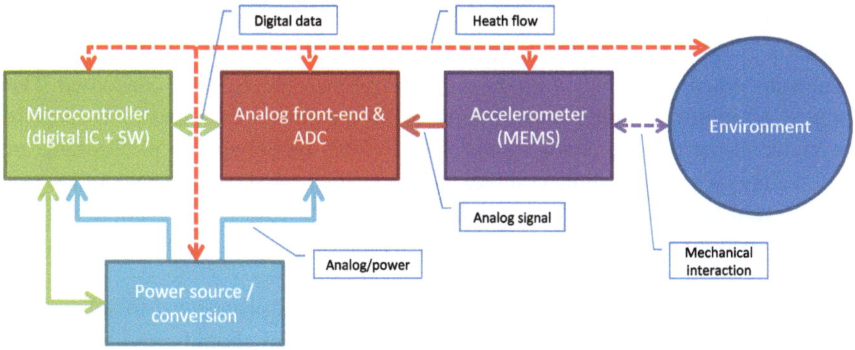


**Fig. 2.8** Example of smart embedded system scenario with interactions among components and with the environment

which may lead to effects such as the increasing signal propagation delay within the microcontroller due to ambient temperature, the rise of temperature of the power conversion elements, and the variation of the accelerometer transfer functions due to heating.

The basic building block of a digital circuit has been a single transistor or standard cell for years, during the rising of the digital era; today a microprocessor core can be instantiated and networked by a designer with the same simplicity with which a flip-flop could be inserted in a design 20 years ago. The main difference is that now this seemingly immediate operation entails even deeper and different physical interactions in the system, and therefore, wider interdisciplinary design basics on one hand, and advanced software tools capabilities on the other [6].

The designer needs to be aware of the description of each interaction, including the nature of the communication, its time scale, and its detail level. The design tool should enable the instantiation and connection of the models of different components, and then the simulation of the entire system. The model of each component should be sufficiently accurate so as to show application-relevant effects but abstract enough so as not to excessively slow down the simulation. Different selectable component models may be useful, with variable levels of detail depending on the application requirements, and with defined validity ranges (e.g., temperature and frequency). The component models to be used, composing a "library" for the system integrator, may have different sources, depending on their specific development flow, but they need to be specified in a shared format readable by the design platform, they may be the output of abstraction flows of lower-level models produced by the device designers; otherwise, the models may be built by the devices suppliers or directly by the system integrators based on datasheets and characterization tests. It should as well be possible to model non-miniaturized, electric or mechanical devices. Roughly speaking, the device model to instantiate should appear as a parameterizable "black box" to conceal design details that cannot be appreciated by the system integrator and to limit the design complexity. In addition to this, it is needed to describe the system assembly, including the board (substrate) material, the distance between components and any contact surfaces between them. These points have a substantial effect on the system behavior, especially when, coherently with the miniaturization trend, SiPs are concerned. This may also require changing some parameters in the device model so as to adapt its behavior to the desired configuration. Hierarchical modeling and partitioning are key features when describing complex systems. The user should be able to describe the relevant features of the use environment together with its interaction with the system, including temperature, RF fields, vibrations, etc., in static and dynamic conditions.

Finally, design constraints need to be described and propagated to subsystems and components in a top–down stream. The design tool should provide different system-level simulation options by trading-off accuracy (varying model accuracy, activation/deactivation of physical models, time scale, etc.) and speed. In this way, initial design space exploration could be easily and quickly performed (maybe using only a single transaction-level simulator engine), while in later stages pre-prototype design validation would be achievable together with a constraints check; if needed co-simulation with logic, circuital, finite element method (FEM) engines, etc., should also be exploitable in a seamless way.

## References

1. I. Akyildiz, W. Su, Y. Sankarasubramaniam, E. Cayirci, Wireless sensor networks: a survey. Elsevier Comput. Netw. **38**(4), 393–422 (2002)
2. D. Barnhart, T. Vladimirova, M. Sweeting, System-on-a-chip design of self-powered wireless sensor nodes for hostile environments, in *Proceedings of IEEE Aerospace Conference*, March 2007, pp. 1–12
3. S. Bergbreiter, K. Pister, Design of an autonomous jumping microrobot, in *Proceedings of IEEE International Conference on Robotics and Automation*, April 2007, pp. 447–453
4. Y. Chuo, M. Marzencki, B. Hung, C. Jaggernauth, K. Tavakolian, P. Lin, B. Kaminska, Mechanically flexible wireless multisensor platform for human physical activity and vitals monitoring. IEEE Trans. Biomed. Circuits Syst. **4**(5), 281–294 (2010)
5. Cooper US, Inc. Energy harvesting (EH) power supply and repeater. (2015). Available: http://www.cooperindustries.com/content/public/en/power_systems/products/automation_and_control/amr_ami.html
6. M. Crepaldi, et al., Towards multi-domain and multi-physical electronic design. IEEE Circuits Syst. Mag. **15**(3), 18–43 (2015)
7. D. Culler, J. Hill, M. Horton, K. Pister, R. Szewczyk, A. Woo, MICA: the commercialization of the microsensor motes. Sensors **19**(4), 40–48 (2012)
8. S. Ergen, A. Sangiovanni-Vincentelli, X. Sun, R. Tebano, S. Alalusi, G. Audisio, M. Sabatini, The tire as an intelligent sensor. IEEE Trans. Comput. Aided Des. Integr. Circuits Syst. **28**(7), 941–955 (2009)
9. Fraunhofer IPMS. MEMS scanners. (2015). Available: http://www.ipms.fraunhofer.de/en/applications/mems-scanners.html
10. A. Gopalai, S. Senanayake, A wearable real-time intelligent posture corrective system using vibrotactile feedback. IEEE/ASME Trans. Mechatron. **16**(5), 827–834 (2011)
11. S. Gowrishankar, T. Basavaraju, D. Manjaiah, S. Sarkar, in *Proceedings World Congress on Engineering*, vol. 1, 2008, pp. 176–187

12. R. Herold, U. Vogel, B. Richter, OLED-on-CMOS based single chip microdisplay and image sensor device, in *Proceedings of International Display Workshop*, 2010, pp. 339–340
13. D. Nosowitz, NFL testing helmets with impact-sensing accelerometers for concussion analysis (2011). Available: http://www.popsci.com/technology/article/2011-01/nfl-test-helmets-impact-sensing-accelerometers-concussion-analysis
14. N.-H. Nguyen, Q.-T. Tran, J.-M. Leger, T.-P. Vuong, A real-time control using wireless sensor network for intelligent energy management system in buildings, in *Proceedings of IEEE Workshop on Environmental Energy and Structural Monitoring Systems (EESMS)*, September 2010, pp. 87–92
15. N.V. Philips , Philips hue. (2015). Available: http://www.meethue.com/
16. D. Reuter, T. Gessner, Smart systems integration by using micro- and nanotechnologies, in *Proceedings of International Multi-Conference on Systems, Signals and Devices*, 2009, pp. 1–8
17. G. Shi, C.-S. Chan, G. Zhang, W. Li, P. Leong, K.-S. Leung, Towards a mobile airbag system using MEMS sensors and embedded intelligence, in *Proceedings of IEEE International Conference on Robotics and Biomimetics*, December 2007, pp. 634–639
18. T. Singh, Hybrid power bracelet runs on solar energy and body heat, inhabitat – sustainable design innovation, eco architecture, green building (2010). Available: http://inhabitat.com/hybrid-power-bracelet-runs-on-solar-and-thermal-energy/
19. Smart Systems in the Multi-Annual Strategic Research and Innovation Agenda of the JTI ECSEL, Part D, EPoSS Industry Association, March 2014
20. N.H. Tan, Tire pressure monitoring system (TPMS) with SP37, in *International All-Terrain Vehicle Symposium*, 2010
21. T. Torfs, V. Leonov, R. Vullers, Pulse oximeter fully powered by human body heat. Sens. Transducers J. **80**(6), 1230–1238 (2007)
22. M. Van Bavel, V. Leonov, R. Yazicioglu, T. Torfs, V. Van Hoof, N. Posthuma, R. Vullers, Wearable battery-free wireless 2-channel EEG systems powered by energy scavengers. Sens. Transducers J. **94**(7), 103–115 (2008)
23. Wispry, Inc. Technology – About MEMS. (2013). Available: http://www.wispry.com/technology.php
24. J. Wolff, T. Heuer, H. Gao, M. Weinmann, S. Voit, U. Hartmann, Parking monitor system based on magnetic field sensor, in *Proceedings of IEEE Intelligent Transportation Systems Conference*, 2006, pp. 1275–1279
25. G.Q. Zhang, A.J van Roosmalen, The changing landscape of micro/nanoelectronics, in *More than Moore: Creating High Value Micro/Nanoelectronics Systems* (Springer, Berlin, 2009)
26. L. Zhou, Y.-A. Chapius, Y. Fukuta, Y. Mita, F. Braun, H. Fujita, Architecture and implementation of distributed control system for MEMS-based intelligent motion surface, in *Proceedings of IEEE International Symposium on Industrial Electronics*, June 2005, vol. 3, pp. 1043–1048

# Chapter 3
# Design Domains and Abstraction Levels for Effective Smart System Simulation

**Sara Vinco, Michele Lora, Valerio Guarnieri, Jan Vanhese, Dimitrios Trachanis, and Franco Fummi**

## 3.1 Introduction

Smart systems represent a broad class of systems defined as intelligent, miniaturized devices incorporating functionality like sensing, actuation, and control. In order to support these functions, they must include sophisticated and heterogeneous components and subsystems such as: application-specific sensors and actuators, multiple power sources and storage devices, intelligence in the form of power management, baseband computation, digital signal processing, power actuators, and subsystems for various types of wireless connectivity (as shown in Fig. 3.1).

Smart components and subsystems are developed and produced with very different technologies and materials specific to the corresponding domain and technology. The heterogeneity involves not only the language or framework adopted, but also different levels of abstraction and different communication and synchronization

---

S. Vinco
Politecnico di Torino, Torino, Italy
e-mail: sara.vinco@polito.it

M. Lora
University of Verona, Verona, Italy
e-mail: michele.lora@univr.it

V. Guarnieri
EDALab s.r.l., Verona, Italy
e-mail: valerio.guarnieri@edalab.it

J. Vanhese • D. Trachanis
Keysight Technologies, Rotselaar, Belgium
e-mail: jan_vanhese@keysight.com; dimitrios.trachanis@keysight.com

F. Fummi (✉)
University of Verona and EDALab s.r.l., Verona, Italy
e-mail: franco.fummi@univr.it; franco.fummi@edalab.it



N. Bombieri et al. (eds.), *Smart Systems Integration and Simulation*,
DOI 10.1007/978-3-319-27392-1_3

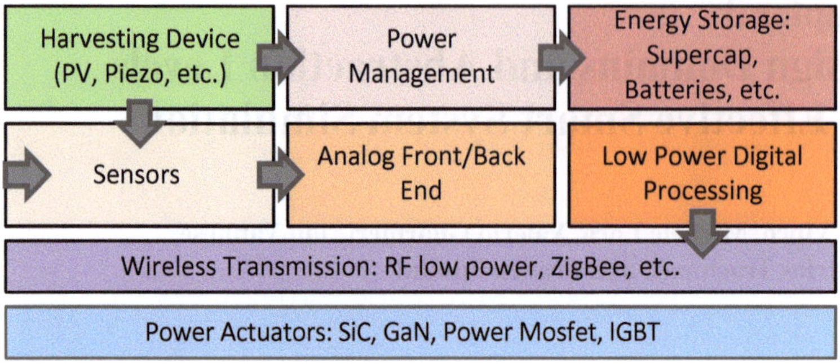


**Fig. 3.1** Typical components of a smart system

styles. As of today, no design methodology and tools exist that can master, simultaneously and in a seamless manner, all the challenges that designers of smart micro-systems are confronted with when new products need to be developed. Nevertheless, modeling and design capabilities for heterogeneous components and subsystems are today available at specialized design houses and silicon makers in various forms. On the other hand, system integrators typically have separate tooling to model the environment. As a result, the challenge in the realization of a smart system goes beyond the design of the individual components and subsystems (an already difficult task by itself), but rather consists in accommodating a multitude of functionality, technologies, and materials.

In this context, simulation is a very critical task, as each component domain adopts specific tools and frameworks, that do not cover the whole smart system heterogeneity. On the other hand, simulation is a key phase in the design and verification process of a system, as it heavily impacts time-to-market and the competitiveness of the final product.

The goal of this chapter is to ease simulation and validation of smart systems. In this context, it provides a *taxonomy of abstraction level/design domains*, to highlight challenges and tools available for each domain. This allows to identify a precise role in the design flow for co-simulation and simulation scenarios, and thus to outline the possible strategies for gaining correct simulation of smart system. The two complementary approaches are compared with the goal of showing respective strengths and weaknesses.

The chapter also aims at enhancing *reuse and integration* by showing how state-of-the-art and commercial tools can ease the adoption of homogeneous simulation, with automatic code generation from lower abstraction levels and automatic integration of heterogeneous interfaces.

As a result, the chapter builds a *comprehensive modeling and simulation framework* that supports digital, analogue, and circuit-level descriptions simultaneously. This improves the contemporary smart systems design flow in such a way that a system-level simulation of all the heterogeneous components/subsystems of a smart system will be possible. This advances state-of-the-art approaches by supporting the development of smart systems, their integration, and efficient simulation.

The chapter is organized as follows. Section 3.2 provides a background on smart system design, by listing available frameworks and formalisms, together with state-of-the-art tools. Section 3.3 identifies the typical abstraction levels and design domains involved in smart system design, with the goal of defining a taxonomy of the most widespread tools and languages. Finally, Sect. 3.4 proposes code conversion and generation approaches to gain homogeneous simulation of the heterogeneous components of a smart system, by working on both language and formalisms. Section 3.5 provides experimental evidence of the proposed solutions, and Sect. 3.6 concludes the chapter with some concluding remarks.

## 3.2 Background on Smart System Modeling

Smart components (and sub-components) are developed and produced with very different technologies and materials, specific to the corresponding domain. The goal of this section is to provide the necessary background for the proposed methodologies. Section 3.2.1 outlines the most widespread formalisms and frameworks available in the literature for tackling smart systems heterogeneity, while Sect. 3.2.2 deepens the ones adopted in the proposed flow.

### *3.2.1 Formalisms and Frameworks for Smart System Modeling*

The heterogeneity of smart system involves not only the language or framework adopted, but also different levels of abstraction and communication and synchronization styles. An evidence of this are the adopted *description languages*, that only target specific domains, such as digital and software components (SpecC and SystemC) or analogue components (VHDL-AMS, Verilog-AMS, and SystemC-AMS).

In the literature, the main approaches proposed for handling such a heterogeneity are (1) top–down flows, relying either on model-based design (MBD) or on models of computation (MoCs), and (2) co-simulation, which exploits different simulation environments to take care of the heterogeneity of the system [13].

In *MBD approaches*, the system model is at the center of the design process and it is continually refined throughout a strictly top–down development flow [5, 22, 26]. Components following different synchronization mechanisms are put together through data conversion, that must be implemented manually, thus not guaranteeing correct integration.

Several *MoCs* have been proposed to describe different aspects of smart systems. As an example, extended finite state machines (EFSMs) [18] are an enhancement of traditional FSMs suited for describing digital HW components and cycle-accurate protocols, while hybrid automata have been defined to allow the integration of continuous physical dynamics with discrete behaviors [20]. Unfortunately, every MoC is a stand-alone environment that cannot cover all the domains comprised in smart system development. Forcing communication through manual conversions between MoCs does not provide any guarantee of correctness of the final result.

The complementary approach is to integrate existing components in a bottom–up flow. This is realized with *co-simulation* environments where each component is simulated in its native environment and framework. Different simulators are then connected by defining rules and conversions about time management and event ordering, supported methods of communication, and rules of process activation [10, 17]. However, co-simulation assemblies heterogeneous components without providing a rigorous formal support, and it only moves the problem of integrating heterogeneous components to the problem of integrating different simulators.

## 3.2.2 Adopted Platforms for Smart System Design

This section gives a very brief overview of the main tools and platforms used in this work for smart system modeling and integration: SystemC and SystemC-AMS as a language supporting a number of abstraction levels (Sect. 3.2.2.1), HIFSuite for automatic conversion of reused code and components (Sect. 3.2.2.2), SystemVue as a simulator and co-simulator (Sect. 3.2.2.3), and UNIVERCM as a MoC spanning across various heterogeneous domains (Sect. 3.2.2.4).

### 3.2.2.1 SystemC and SystemC-AMS

SystemC is a widely deployed extension to C/C++ for describing HW constructs, ranging from register-transfer level up to transactional level [1]. The underlying simulation kernel is entirely event-based, i.e., a centralized scheduler controls the execution of processes based on events (synchronization, time notifications, or signal value changes). SystemC provides also a methodology for performing abstract modeling, simulation through generalized modeling of communication and synchronization: transaction level modeling (TLM) [3].

The SystemC simulation kernel has not been natively designed to handle the modeling and simulation of analog/continuous time systems. The recent extension SystemC-AMS [2] was designed for overcoming this lack, i.e., for modeling and simulating interacting analog/mixed signal functional subsystems, thus allowing to extend the adoption of a SystemC-based environment also to extra-functional, continuous time domains.

SystemC-AMS provides different abstraction levels to cover a wide variety of domains. *Timed data-flow* (TDF) models discrete time processes, that are scheduled statically by considering their producer–consumer dependencies. *Linear signal flow* (LSF) supports the modeling of continuous time behaviors through a library of predefined primitive modules (e.g., integration, or delay), each associated with

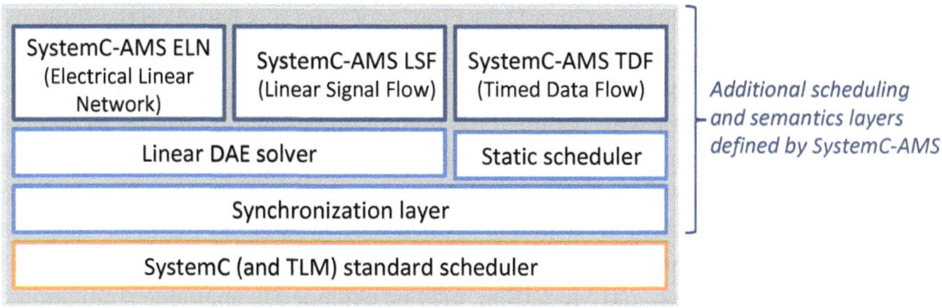


**Fig. 3.2** Architecture of the SystemC scheduler as extended with SystemC-AMS support

a linear equation. *Electrical linear network* (ELN) models electrical networks through the instantiation of predefined primitives, e.g., resistors or capacitors, where each primitive is associated with an electrical equation. In case of ELN or LSF descriptions, a SystemC-AMS AD solver analyzes the ELN and LSF components to derive the equations modeling system behavior, that are solved to determine system state at any simulation time.

As highlighted in Fig. 3.2, the key feature of all SystemC extensions is that overall simulation is handled by the sole SystemC simulation kernel, that interacts with its extensions to define, time after time, both the execution queue and the corresponding system evolution.

#### 3.2.2.2 HIFSuite

HIFSuite is a set of tools and application programming interfaces (APIs) that provide support for modeling and verification of HW/SW systems [15]. The core of HIFSuite is the HDL Intermediate Format (HIF) language upon which a set of front-end and back-end tools have been developed to allow the conversion of HDL code into HIF code and vice versa. HIFSuite allows designers to manipulate and integrate heterogeneous components implemented by using different hardware description languages (HDLs). Moreover, HIFSuite includes tools, which rely on HIF APIs, for manipulating HIF descriptions in order to support code abstraction/refinement and post-refinement verification, including $A^2T$, a tool for abstracting RTL digital components to TLM or C++ [8].

#### 3.2.2.3 SystemVue

SystemVue is an electronic design automation (EDA) environment for electronic system-level (ESL) design, focused on RF and DSP systems [4]. It supports complex RF envelope carriers and dataflow simulations [21]. In SystemVue, a system is described as a schematic of components connected with wires and busses. The simulation technology is based on a Data-Flow MoC and it is based on the Ptolemy multi-domain, heterogeneous simulation platform [22].

SystemVue is well suited for the integration of heterogeneous systems. It provides numerous libraries with parameterized components and interfaces to diverse modeling formats, ranging from MATLAB to the main HDLs, such as Verilog and VHDL. Furthermore, it allows to create custom components in math language or C++ and to add them to a purely SystemVue system. SystemVue supports multi-domain simulations through links to event-based as well as circuit simulation engines, such as SystemC and ModelSim, may be extended to analogue simulations.

#### 3.2.2.4 UNIVERCM

UNIVERCM is an automaton-based formalism that unifies the modeling of both the analogue (i.e., continuous) and the digital (i.e., discrete) domains, as well as hardware-dependent SW. A formal and complete definition is available in [19].

In each UNIVERCM automaton (depicted in Fig. 3.3), states model the continuous dynamics of the system as a condition that must be satisfied to perform continuous evolution (invariant) and a predicate modeling the evolution of variables over time (flow). Edges between states model the discrete dynamics as evolution of variables and activation of synchronization events, controlled by a boolean predicate on the variable state and by synchronization checks.

UNIVERCM is an important resource in smart system design as it is well suited for the application to heterogeneous domains [19]. Indeed, the computational model allows to cover the heterogeneity that characterizes such systems, ranging from analogue and digital HW up to dedicated SW. Guglielmo et al. [19] presented a comprehensive reuse and design flow based on UNIVERCM, thus showing how it is possible to provide formal rules and automatic tools to convert the heterogeneity to UNIVERCM and to produce a homogeneous simulatable implementation of the generated UNIVERCM system. Thus, UNIVERCM enhances reuse and bottom–up design.

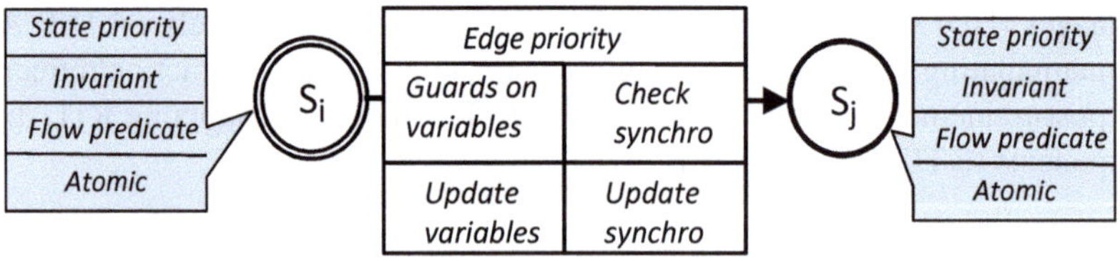


**Fig. 3.3** Example of UNIVERCM automaton

## 3.3 Analysis of Smart System Simulation Solutions

Simulation and design are heavily influenced by the abstraction level of each component and, as a consequence, by the level of heterogeneity that characterizes the system in terms of domains, abstraction levels, and synchronization mechanisms. It is thus necessary to clearly identify the abstraction level involved in smart system design (Sect. 3.3.2) and to associate each domain and simulator to the correct level. For this reason, this section proposes a taxonomy that associates frameworks and design flows to each domain and abstraction level (Sect. 3.3.3). This constitutes a necessary starting point for understanding the impact of abstraction levels and of the heterogeneity/homogeneity trade-off on simulation (Sect. 3.3.4).

### *3.3.1 Typical Domains of Smart System Design*

The typical classes of components of any smart system are identified in terms both of constituting characteristics and of role *w.r.t.* the inner information/energy flows. For this reason, components are sub-divided into six main domains:

- *MEMS, sensors, and actuators*, in charge of communicating with the surrounding environment;
- *Power sources*, necessary to guarantee correct functioning of all other components;
- *Discrete and power devices*, as parts of the energy flow, responsible for energy dispatching and harvesting;
- *Analogue and RF components*, mainly responsible for signal processing, transmission, and reception;
- *Digital HW*, core of the system processing and functionality;
- *Embedded SW*, as system controller and main mean of communication with the end users.

The main simulation problems of smart systems derive from this heterogeneity that requires the use of different design languages and different abstraction levels. Moreover, it is extremely unlikely that a single team has the knowledge to cover all such design domains, thus, we have to assume that a set of design teams must cooperate by using their own favorite design languages. In fact, there is no *Esperanto* able to effectively model all such domains. A variety of design languages has rather been proposed in the past decades to cover specific design domains, and some standards de facto became the reference languages for design teams specialized in each design domain. This challenging scenario will be the focus of the next sections.

### 3.3.2 Abstraction Levels of Smart System Design

The main factors determining the level of abstraction are: time granularity, interconnection model, state space granularity, and data aggregation. *Time granularity* is an important dimension in a heterogeneous environment. It may be continuous or discrete time, or follow an event-based semantics where time ticks only when the system state changes. The *interconnection model* describes communication and synchronization between components as potential or flow quantities (conservative systems), flow charts, or transactions. The *granularity of state space* details data aggregation for simulation purposes, i.e., variables managed by differential equations, symbolic variables, or objective constructs (i.e., system state describes the possible behavior, e.g., C++). Finally, *data aggregation* states whether the component is modeled by considering the minimum (black box) or maximum (clear box) number of state space variables necessary for a correct representation of the observable behavior.

Given these factors, it is possible to identify five main abstraction levels, typical of smart systems.

- At *transactional* level, simulation is strictly event-based and inter-component communication happens via transactions (that provide a communication protocol to the system). System state is modeled with variables.
- At *functional* level, simulation is event-based but communication relies on the flow chart interconnection style.
- The *structural* level has two main approaches depending on time granularity. Continuous time evolution is modeled with differential equations and by observing conservative laws. Discrete time may adopt both event-based or flow chart synchronization, and finite set variables are adopted.
- At *device* level, simulation can be both continuous or discrete time. The major difference is that at device level all variables are modeled explicitly, while structural level models only those variables that are strictly necessary for simulation purposes.
- The *physical* level adopts continuous time synchronization and the conservative interconnection style. State space is described with continuous fields as differential equations and all variables are modeled in a clear box approach.

### 3.3.3 Design-Domains/Simulation-Level Taxonomy

Given the variety of abstraction levels and the heterogeneous domains typically present in any smart system, it is possible to build the design-domains/simulation-level taxonomy shown in Fig. 3.4. Such a chart identifies the abstraction level (rows) and the domain (column) of the most widespread tool and languages adopted in the context of smart systems. This allows to correctly differentiate the use of co-simulation and simulation according to the two dimensions. Text in bold shows the typical entrance level and tools for each domain.

| | *MEMS, sensors and actuators* | *Power sources* | *Discrete and power devices* | *Analog and RF* | *Digital HW* | *Embedded SW* | |
|---|---|---|---|---|---|---|---|
| TRANSACTIONAL | SystemVue | SystemVue | SystemVue | SystemVue | SystemVue, SystemC (TLM/AMS) | SystemVue, QEMU | SIMULATION |
| FUNCTIONAL | C++ | C++ | C++ | C++, SystemVue | C++, SystemC | **C++, QEMU** | SIMULATION |
| STRUCTURAL | ADS, Matlab, AMS HDLs, MEMS+ | Matlab, Simulink, AMS HDLs | ADS | ADS, Matlab, AMS HDLs | **RTL HDL** | Cycle accurate QEMU | CO-SIMULATION |
| DEVICE | Matlab, MEMS+, AMS HDLs | FEM, Spice | EMPro, Spectre, Momentum | EMPro, Spectre, Momentum | AMS HDLs | – | CO-SIMULATION |
| PHYSICAL | **Matlab, MEMS+, FEM** | **FEM, Spice** | **EMPro, Spectre, Momentum** | **EMPro, Spectre, Momentum** | AMS HDLs | – | CO-SIMULATION |

**Fig. 3.4** Design-domains/simulation-level taxonomy, identifying the abstraction level (*rows*) and the domain (*column*) of the most widespread tool and languages adopted in the context of smart systems. *Text in bold* shows the typical entrance level and tools for each domain

Models belonging the *lowest abstraction levels* (i.e., physical, device, and structural) are represented by different domain-specific design languages. They must thus be simulated by using their own simulator (e.g., Matlab, Modelsim, EMPro). For this reason, a framework covering more than one domain can be implemented only by using *co-simulation techniques* which connect different tools by exchanging simulation data from one tool to another.

Moving to the *functional level*, there is a convergence in the modeling language, as all models belonging to different domains are represented in C++. This would in principle allow a simulation among different domains. However, the MoC implemented into each C++ model can be different from domain to domain. Thus, simulation cannot be simply obtained by linking functional C++ models, but such models must also be coherent *w.r.t.* the same MoC. Thus, either the chosen MoC covers all domains or some data and synchronization conversion is necessary.

At *transaction level*, simulation frameworks enforce a common transaction-based communication protocol to all domains. This allows to seamlessly integrate components belonging to different domains and based on different MoCs and synchronization mechanisms.

### 3.3.4 Impact of MoCs on Simulation and Co-simulation Performance

The taxonomy in Fig. 3.4 helps in further understanding the impact of MoCs and of heterogeneity on simulation and co-simulation at different abstraction levels.

As mentioned in Sect. 3.3, the heterogeneity of the *lowest abstraction levels* forces to simulate each design domain by using ad-hoc simulators. Co-simulation frameworks are thus built by connecting different simulators, such as shown in [10, 17]. Unfortunately, explicitly modeling the synchronization between simulators, different for language, formalism, and underlying MoC, heavily impacts simulation performance and effectiveness [19]. Other approaches achieve a lighter impact by compiling separately the different formats and linking them together, such as done by ModelSim to co-simulate SystemC and VHDL. This lighter approach is still affected by the presence of heterogeneous MoCs, as the data sharing mechanism and time synchronization introduce a heavy overhead.

*Functional level* brings to a convergence in terms of modeling language and framework, thus showing the impact of MoCs to the full. If all C++ components follow the same MoC, then they can easily be integrated with no further overhead. Else, if the adopted MoCs are heterogeneous, it becomes necessary to introduce a communication layer for applying data and synchronization conversion.

Communication and synchronization are further eased at *transaction level*, as transactions and standard interfaces force a single communication protocol to all components. This mitigates the effect of having multiple MoCs, as problems risen by data sharing and time synchronization are moved inside the transactional communication mechanism.

This analysis highlights that the heterogeneity of smart systems impacts simulation performance in many directions. Contributing elements are indeed the adopted languages, the levels of abstraction, and the MoCs followed by the components to be integrated. The weakest approach appears to be co-simulation, mandatory at lowest levels, as it pays the price of all degrees of heterogeneity. Simulation becomes more effective at functional and transactional levels, where heterogeneity is constrained and limited to few synchronization mechanisms. For these reasons, the remainder of this chapter will focus on code generation for effective simulation of smart systems at functional and transactional levels.

## 3.4 Proposed Methodologies

The analysis of the smart system simulation scenarios proposed in the previous section highlighted that the choices in terms of abstraction level, language, and MoC may heavily affect simulation performance. This section outlines three alternatives, different in terms of implementation choices and covered domains. The solutions are summarized in Fig. 3.5, and they provide different coverage/performance trade-offs, together with techniques and tools for achieving automatic generation of simulatable code. Section 3.4.1 focuses on functional level, and it estimates the impact of MoCs on simulation. On the other hand, Sect. 3.4.2 provides two solutions at transactional level, based on SystemC and on the SystemVue framework.

### *3.4.1 Smart System Simulation at Functional Level*

The functional level brings all domains to a convergence in terms of modeling language, usually C++. This easies the achievement of simultaneous simulation of components belonging to different domains. At the same time, an effort may

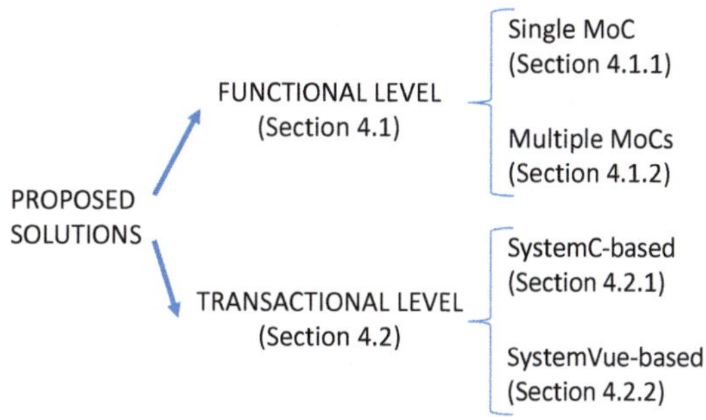


**Fig. 3.5** Proposed solutions for homogeneous simulation of heterogeneous smart systems

be necessary whenever the C++ representations of components follow different MoCs, i.e., different synchronization management rules. This section provides an example for both flows, with the goal of showing the impact of MoCs to the full.

#### 3.4.1.1 Simulation Based on a Single MoC

The UNIVERCM MoC, presented in Sect. 3.2.2.4, was designed to reconcile heterogeneous domains to a unique formalism. It supports a full bottom–up approach where already existing heterogeneous descriptions can be automatically converted and integrated into UNIVERCM automata for being, subsequently, re-mapped to a single simulatable model. This section details both the flows, with a focus on the major conversion issues and solutions.

##### Mapping from Heterogeneity to UNIVERCM

The strategy to map any component to UNIVERCM strictly depends on the domain and abstraction level of the starting description [14].

Mapping *digital HW descriptions* in UNIVERCM requires to reproduce the simulation semantics of HDLs, both in terms of scheduling and of synchronization.

HDL processes are represented as automata. All edges of an automaton are guarded by the activation of synchronization labels, reproducing a value change of any of the signals in the sensitivity list. This activates an automaton in response to changes in its sensitivity list. Note that the propagation of synchronization events is straightforward, as labels are instantaneously visible from any automaton.

The typical HDL scheduling routine is in charge of generating and propagating events and advancing simulation time. This mechanism must be represented in UNIVERCM so that events are processed in the same order and simulation semantics is preserved. The main feature that must be preserved is thus the fact that simulation time is advanced only when there is no event to be processed in the system nor any signal to be updated. The scheduling routine is represented with an additional automaton, that advances a continuous variable representing time only when there is no active label in the system. This allows to process events in the same order as in the original HDL and to preserve the original simulation semantics.

*HW-dependent SW* (HdS) is SW that controls and abstracts HW functionality, to allow easy and standard access to HW devices and the deployment of more abstracted SW. HdS is thus in charge of managing communication with HW and it needs to be reactive to signals and interrupts risen by HW devices. Each HdS function is mapped to a UNIVERCM automaton, evolving among a certain set of states via transitions (note that continuous time evolution is not supported for this domain). Each function is provided with two special labels: an activation label (representing function invocation and activated by automata willing to execute the function) and a return label (used to communicate to the caller that the function has finished its execution). This allows inter-function communication. Automata representing HdS functions can be also sensitive to events coming from HW automata, representing HW interrupts. This, together with data sharing for modeling MMIO mechanisms, allows to reproduce the basics of HW–SW communication. An example of HW–SW communication, and of mapping to UNIVERCM of the corresponding components, is provided in Fig. 3.6.

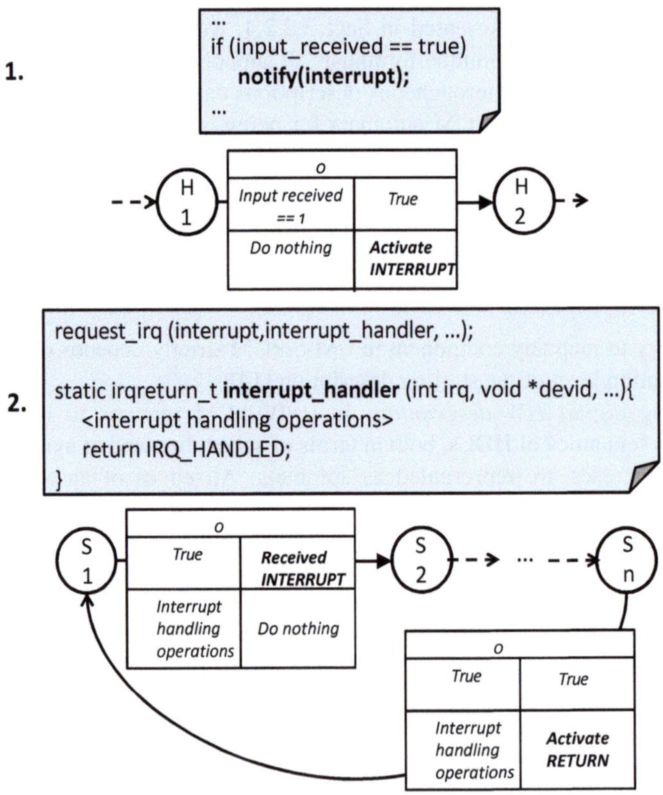


**Fig. 3.6** Mapping to UNIVERCM of a digital HW component firing an interrupt (*1*) and of the corresponding interrupt service routine (2)

UNIVERCM can be easily adopted to model also *analogue models* described with differential equations as hybrid automata [16, 27]. The mapping is straightforward, even if some transformations are necessary to reproduce the synchronization semantics and to remove hierarchy from the automata.

Once that all starting descriptions have been converted to UNIVERCM, automata evolve simultaneously through data sharing (i.e., by accessing the same variables) and by synchronizing via labels. Thus, no additional communication or scheduling mechanism is necessary.

### Mapping from UNIVERCM to C++

The conversion flow from UNIVERCM to C++ is defined in general for any automata, with no concern regarding the language of the original description converter to UNIVERCM.

Each UNIVERCM automaton is mapped to a C++ function, representing the whole automaton evolution, as depicted in Fig. 3.7. A state variable is used to store the current state of the automaton. The function body is built as a `switch` statement, where each case represents one of the automaton states. Each state case lists the implementation of all the outgoing edges and of the delay transition provided for the state.

Each *edge* is implemented as an `if` or `else if` statement, whose guard is a logic and of the enabling condition on the edge and of the activation condition on synchronization events. The body executed when the guard is satisfied includes the update of variables and the activation of synchronization events. Furthermore, the state variable is updated to the destination state of the edge.

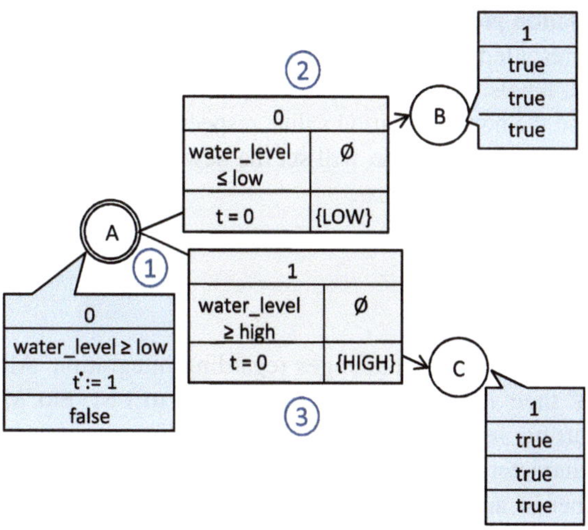


```
switch(state) {
 case (A):
   if(water_level ≥ Xhigh){
     t := 0;
     state := C;          (3)
     activate(HIGH);
   }
   else if(water_level ≥ Xlow){
     ṫ := 1;
     state := A;          (1)
   }
   else if(water_level ≤ Xlow){
     t := 0;
     state := B;          (2)
     activate(LOW);
   }
 case (B): ...
 case (C): ...
}
```

**Fig. 3.7** UNIVERCM automaton to be converted to C++ (*left*) and corresponding generated code (*right*)

*Continuous evolution* is implemented as an `if` or `else if` statement whose guard is the invariant condition that allows to remain in the state. The body executed when the guard is true implements a discretized implementation of the flow predicate, by adopting the *Euler numerical integration algorithm* with time discretization step chosen by the designer [11]. It is important to note that the Euler method can be replaced with one of the many available algorithms for the approximation of solutions of ordinary differential equations.

Code generated from UNIVERCM automata is ruled by a *management function*, in charge of activating automata and of managing the status of the overall system and parallel composition of automata. The result of this approach is that all code generated from UNIVERCM automata is controlled by a single function, and it is thus provided with a simple interface.

If the system is made of more UNIVERCM automata, the management function is not enough to grant correct composition. Indeed, the starting components come from heterogeneous domains, and thus the communication means may differ. On the other hand, communication between UNIVERCM automata happens via variable sharing and through synchronization events. Thus, any two automata can be easily composed by checking the correspondence between variables and synchronization events of the two. Mapping the one in the other must be identified by the designer. This allows to extend the management function to all operations necessary to propagate updated values.

Finally, UNIVERCM variables and events are mapped to native C++ constructs. Variables are mapped to a couple of C++ variables, representing the current value and the future value, respectively, in order to respect the UNIVERCM semantics. Value update is performed by the management function, as previously anticipated. The type of each variable is determined by the variable alphabet for discrete variables, while continuous variables are mapped to doubles. Support type libraries may be used, for simulation purposes or to enhance simulation speed [9]. Synchronization events are represented with boolean values, where `true` states that the label is active. In detail, labels are mapped to a couple of boolean values, representing the current value and the next simulation value, respectively. At the end of each simulation step, the management function will set the new current value to the future one, and reset the future value to `false`.

### Integration Strategies and Challenges

Simulation based on a single MoC poses no challenges regarding integration. All starting components, despite of their heterogeneity, are converted to UNIVERCM automata, by mapping the starting semantics to UNIVERCM native constructs. This allows to abstract the characteristics of the starting descriptions, and to represent the system as a number of automata that interact through no conversion mechanism. This is a winning approach, as no manual intervention is necessary to allow integration. This reduces by far communication overheads, and it speeds up simulation.

#### 3.4.1.2 Simulation Based on Multiple MoCs

UNIVERCM is a very powerful MoC, as it covers a wide number of domains. However, its representation of digital HW may lead to an explosion of the modeled automata, both in terms of states and of synchronization labels. Furthermore, no methodology has been defined yet for mapping circuit-based descriptions, as electrical behaviors and conservation laws are difficult to reproduce in an automata based approach. For this reason, it may be necessary to integrate code generated via UNIVERCM with C++ code generated with other strategies. This section outlines two additional strategies, necessary to cover all smart system domains efficiently. The section ends by presenting the integration strategies and challenges, to allow overall smart system simulation even in presence of different MoCs.

##### HIFSuite for Efficient Conversion of Digital HW to C++

HIFSuite (introduced in Sect. 3.2.2.2) is a closely integrated set of tools and APIs for reusing already developed components and for verifying their integration into new designs [15].

HIFSuite was first designed for allowing system designers to convert HW/SW design descriptions from a HDL to a different HDL and to manipulate them in a uniform and efficient way. For this reason, the underlying HIF core language is made of a set of objects corresponding to traditional HDL constructs like, for example, processes, variable/signal declarations, sequential and concurrent statements, and so forth [6]. Each HIF construct is mapped to a C++ class that describes specific properties and attributes of the corresponding HDL construct. Such objects can then be manipulated through powerful C++ APIs which allow to explore, manipulate, and extract information from HIF descriptions.

All such characteristics make HIFSuite a very convenient infrastructure to define conversion tools working on digital HW descriptions. The typical conversion flow from digital HW to C++ is outlined in Fig. 3.8, and it leaves the underlying MoC of the starting description unchanged.

Any digital HW description, implemented in a HDL language, is converted to its HIF representation via the *HIFSuite front-end tools*, performing a straightforward mapping from HDL constructs to the corresponding HIF objects. The abstraction of the HIF description is then carried out by two manipulation tools from HIFSuite, DDT and $A^2T$. DDT replaces the original HDL data types from the starting HW description with C++ built-in data types in order to greatly improve simulation performance. Then, $A^2T$ implements the methodology in [7] to convert the HDL processes to functions and the HDL scheduling semantics to a management function. Additionally, $A^2T$ can be guided to generate more performing C++ code by providing it with profiling information of the starting HDL implementation. If the repeated execution of asynchronous processes dominates execution time, $A^2T$ may replace the standard dynamic HDL simulation semantics with a static scheduling approach. Such an approach creates a sequence of processes to be repeated at every

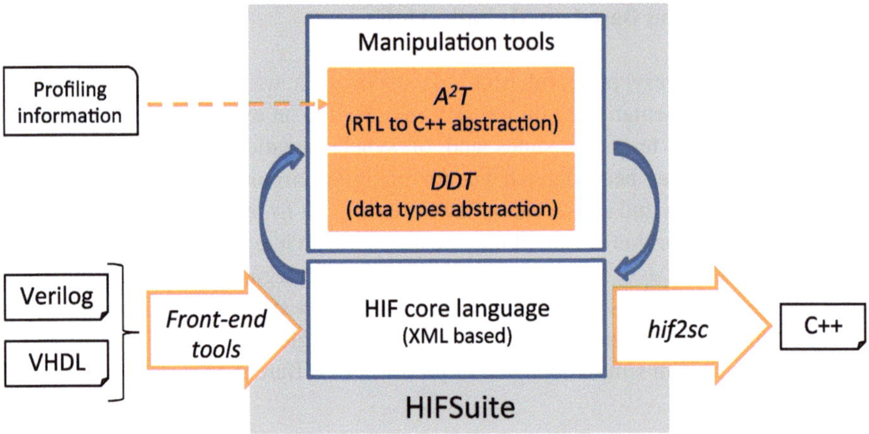


**Fig. 3.8** HIFSuite-based flow for automatic conversion of digital HW descriptions to C++

simulation cycle, thus avoiding the overhead of event management. This allows to *further abstract the starting HDL description*, to customize the generated code with the goal of optimizing simulation performance. The obtained HIF description is finally converted to C++ through the back-end tool hif2sc.

The winning aspect of this strategy *w.r.t.* the UNIVERCM-based conversion flow presented in Sect. 3.4.1.1 lies in the efficiency of the generated code. HIF natively preserves the HDL semantics, thus not introducing additional constructs, e.g., for scheduling or synchronization management. This results in a more compact C++ implementation of the starting digital HW.

#### Conversion of Analogue and Mixed Signal Descriptions to C++

Analogue components can be seen as a set of algebraic and differential equations, expressing the functionality. These equations can be expressed in different ways: they can be explicitly listed or they can be hidden by expressing them as interconnections of primitives, as for block diagrams. Thus, when aiming at reproducing the behavior of an analogue device, it is fundamental to extract the correct set of equations from the original description. To accomplish this task, HIFSuite analysis features come in handy, and they are exploited into a framework of front-end, manipulation and back-end tools. The resulting flow is depicted in Fig. 3.9.

To read analogue descriptions, the Verilog parser of HIFSuite is extended to support Verilog-AMS. The tool takes care of parsing analogue descriptions, based on dipole equations, and to map constructs into HIF. The HIF representation is then used to analyze and manipulate the information expressed by the design. Analysis and manipulation are performed by OCCAM (*Ordinary C++ Code for Analogue Models*), a tool developed on top of HIFSuite that implements an analysis and manipulation algorithm composed by the following five steps:

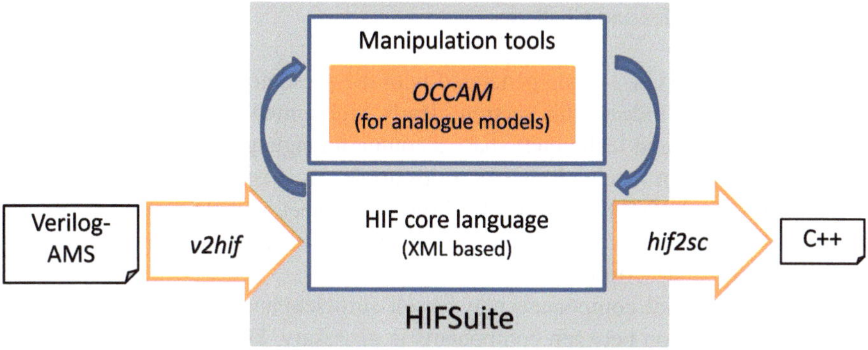


**Fig. 3.9** HIFSuite-based flow for automatic conversion of analogue and mixed signal descriptions to C++

- *Acquisition*: Starting from the set of dipole equations acquired by the Verilog front-end tool, a hash table is created. For each electric branch of the circuit represented by the original description, current and voltage are labeled and then, every equation is stored in the hash table, using the left value label as key. Then, also the inverse equations are computed, stored in the table, and marked as "linearly linked" to the original equation.
- *Enrichment*: The system of equations can be partially specified, and some relations may thus be left implicit. It is necessary to apply Kirchhoff's current and voltage laws to retrieve the entire set of equations composing the system. This is done by employing a modified nodal analysis algorithm on the set of equations extracted during the acquisition step. The implicit equations, retrieved by the modified nodal analysis, are inserted into the hash table and marked as "linearly linked."
- *Assemble*: In order to abstract the system, the outputs of interest are fixed by the designer. For every output of interest, its label is used to fetch an equation from the hash table. Then, all the terms of the fetched equations are used as label to fetch other equations, recursively, in order to retrieve all the terms influencing the chosen output. A tree structure representing these dependencies is built for every output of interest.
- *Preparation*: The tree built at the assemble step is visited, and the dependencies are mapped into a sequence of assignments and function calls, to represent algebraic and differential operators.
- *Dismantle*: The sequence of instructions created after the previous steps are inserted into a function. Since the produced models aim at simulating continuous time evolution, they have to be repeatedly executed. Thus, the simulation scheduler will provide to call and execute the function wrapping the behavior, periodically during the simulation.

Finally, the behavioral representation produced by OCCAM and modeled in HIF has to be translated into C++. To do this, the HIFSuite back-end tools have been extended in order to support this kind of representation, to produce C++ code for the simulation.

Integration Strategies and Challenges

The integration of C++ code generated with the presented techniques introduces major challenges. Indeed, this section clearly highlighted that at functional level different domains and techniques share a common language, but not the MoC and the synchronization mechanisms. As an example, an event fired by a component generated through UNIVERCM may be difficult to detect by code generated through HIFSuite or through a complex abstraction process, necessary to handle analogue and mixed signal descriptions.

If execution inside components may be self-sufficient and correct, problems arise whenever interaction between components is necessary. Due to the complexity of the task and to the complex configurations that may show up, this task can be handled only manually, by carefully considering the characteristics of the specific components into play.

Whenever integrating heterogeneous C++ code, the designer shall consider:

- *Functionality activation*: Each MoC introduces different scheduling strategies in the C++ code, ranging from the reconstruction of HDL scheduling up to simple activation of all automata for UNIVERCM-based code. The designer shall implement a global scheduling routine, that activates the single domains by respecting timing and causality relationships;
- *Time evolution*: Each MoC advances time with specific solutions, that are affected by the presence of runnable activities. Local scheduling strategies must thus agree on a shared notion of time, so that events are propagated in the correct order and that digital synchronous signals such as clocks are coherent *w.r.t.* the remainder of the system;
- *event propagation*: Each local scheduler must be able to detect synchronization events fired by the other domains. For this reason, the global scheduler must convert events from one formalism to the other, without introducing delays or timing misalignments;
- *Data sharing*: Different components must be able to share data despite of the implementation differences. The global scheduling routine shall propagate value changes, thus converting data from one format (or data type) to the other.

This highlights that, even if the single conversion techniques are correct, interaction of heterogeneous MoC introduces heavy management overheads and it may leave space for synchronization misalignments.

### *3.4.2 Smart System Simulation at Transactional Level*

The transactional layer brings all domains to a convergence in terms of modeling language and of underlying framework. The differences in terms of MoC or abstraction level are not reduced by means of conversion methodologies, but they are rather preserved to ease the integration process. Ad-hoc interfaces or simulation strategies mask this heterogeneity with a transaction-based mechanism, where a global scheduler satisfies activation requests and performs all conversions and synchronization with no intervention from the user. This section provides two examples of this strategy, the one relying on the standard language SystemC (Sect. 3.4.2.1) and the other based on the commercial tool SystemVue (Sect. 3.4.2.2). This will highlight the characteristics of the transactional level to the full.

#### 3.4.2.1 SystemC-Based Simulation

SystemC, together with its extensions, is a well-established language for the modeling of smart systems. Its strength, as anticipated in Sect. 3.2.2.1, is the presence of a single simulation kernel, mastering requests coming from any of the supported MoCs and libraries.

SystemC can be considered transactional as any of the supported MoCs defines a precise interface to the simulation kernel, thus wrapping different levels of abstraction of the instantiated constructs. Each solver communicates with the simulation kernel through transactions, i.e., activation requests that are satisfied by the kernel through synchronization with the remainder of the system and through data sharing and conversion. This section shows how effective SystemC can be at supporting the heterogeneity of smart systems, ranging from analogue and mixed signal conservative descriptions up to digital HW components.

##### Mapping from UNIVERCM to SystemC

Mapping of UNIVERCM to SystemC traces the approach for C++ code generation proposed in Sect. 3.4.1. However, the presence of a simulation kernel allows to delegate some management tasks, and to reproduce automata behavior through native SystemC constructs. Note that this is crucial to ease and enhance the interaction with SystemC code generated through different design flows.

The main effect of the adoption of SystemC is on the management routine. UNIVERCM automata are indeed mapped to processes, rather than functions. This allows to delegate automata activation to the SystemC scheduler, by making each process sensitive to its input variables. Automata activation is removed from the management function, that still updates the status of variables and events at any simulation cycle. The management function itself is declared as a process, activated with a custom event after all automata have performed one simulation step.

The mapping of synchronization events is left unchanged, despite of the presence of native SystemC events, i.e., `sc_events`. Indeed, SystemC events cannot be used into conditions, while this is a feature necessary to fully support UNIVERCM transition semantics.

The mapping of UNIVERCM variables changes slightly. Variables shared by two or more automata are mapped to SystemC signals, to allow data sharing between processes and ensure correct simulation and process activation. UNIVERCM variables used by a single automata are still mapped to a couple of C++ variables, i.e., current value and future value, that are updated and handled by the management function.

Mapping of Digital HW to SystemC and SystemC TLM Through HIFSuite

As previously stated in Sects. 3.2.2.2 and 3.4.1.2, HIFSuite is an ideal framework to convert digital HW descriptions into corresponding SystemC and SystemC-TLM descriptions. The flow to automatically convert digital HW descriptions to SystemC at RTL is depicted in Fig. 3.10. The input HW description, written in VHDL or Verilog, is firstly converted to its HIF representation by the *HIFSuite front-end tools*. This step is achieved by parsing the input description and mapping HDL constructs to corresponding HIF objects. Then, the HIF description is converted to the corresponding SystemC RTL code by the back-end tool hif2sc. A number of manipulations on the HIF description are required during this step to account for the lack of expressiveness of SystemC *w.r.t.* VHDL and Verilog. In fact, some VHDL and Verilog constructs do not have a direct mapping to a corresponding SystemC construct. As such, they must be translated by resorting to an equivalent implementation through other SystemC constructs.

HIFSuite also features a flow to automatically abstract digital HW descriptions to SystemC TLM for faster simulation speed. The resulting flow is illustrated in Fig. 3.11. The first step consists again of converting the input HW description to its corresponding HIF representation by the *HIFSuite front-end tools*. If the target is to generate a TLM description optimized for simulation performance, the following step consists of invoking DDT from HIFSuite on the generated HIF description in order to improve simulation performance by replacing the original HDL data types with C++ built-in data types. This step is however completely optional. In case it is bypassed, the output TLM description at the end of the flow will feature SystemC data types. The abstraction of the HIF description from RTL to TLM is carried out by the manipulation of $A^2T$ from HIFSuite. $A^2T$ produces code compliant with the TLM-2.0 standard. The user can select which TLM protocol will be generated by adopting one of the two TLM-2.0 coding styles, namely *loosely timed* (LT) and *approximately timed* (AT). If the LT coding style is adopted, the

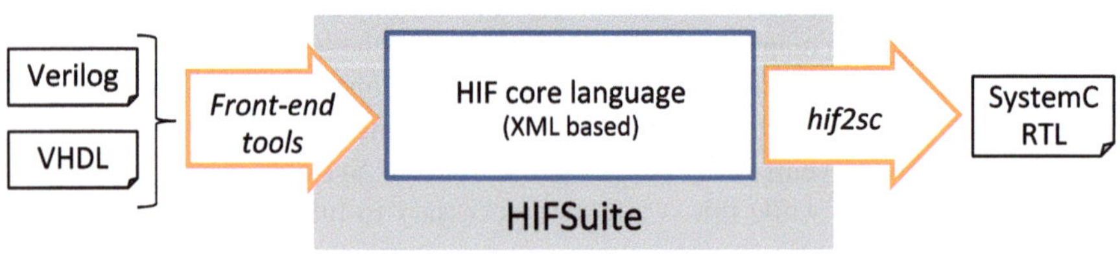


**Fig. 3.10** HIFSuite-based flow for automatic conversion of digital HW descriptions to SystemC RTL

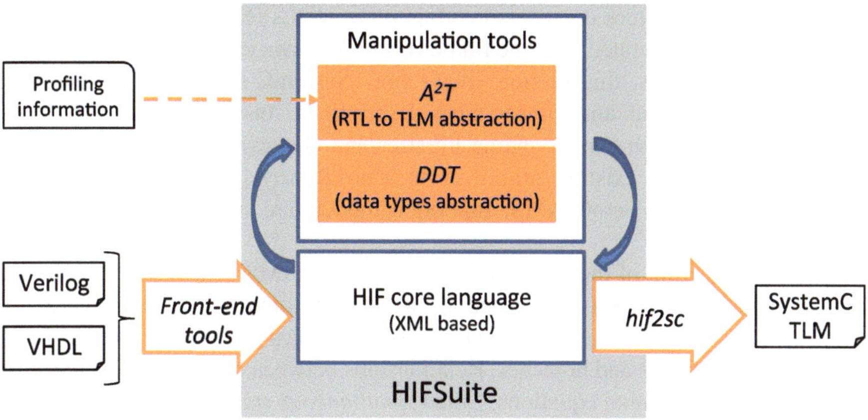


**Fig. 3.11** HIFSuite-based flow for automatic conversion of digital HW descriptions to SystemC TLM

abstracted design will implement the blocking transport interface, and blocking transport primitives will be used to achieve communication. Conversely, if the AT coding style is adopted, the abstracted design will implement the non-blocking transport interface, and non-blocking transport primitives will be used to achieve communication. The abstraction process generates C++ functionality code from RTL processes, and replaces the RTL cycle-accurate communication protocol with the transaction-based TLM communication protocol. As reported in Sect. 3.4.1.2, profiling information on the starting HW description can be provided to $A^2T$ in order to generate more efficient C++ code for the design functionality. Finally, the abstracted HIF description is converted to SystemC TLM through the back-end tool hif2sc.

#### Mapping of Analogue Conservative Descriptions to SystemC-AMS

Smart systems often feature heterogeneous components that do not match the traditional digital design flow. A typical example is MEMS components, often used as means of sensing and actuation, thus having a crucial role in the interaction of the system with the surrounding environment. The main complexity introduced by this kind of descriptions is that they are *both behavioral and conservative*, i.e., they feature a certain level of abstraction *w.r.t.* the actual component realization, but at the same time they obey physical laws, such as energy conservation laws [12, 25].

The limitations of traditional flows and tools at handling such components are highlighted by the characteristics of SystemC-AMS that, though being the reference language for smart system simulation, does not support descriptions that are both behavioral and conservative (as described in Sect. 3.2.2.1). The limited flexibility of SystemC-AMS forces designers to adopt other HDLs (e.g., Verilog-AMS), that cannot be easily integrated with the frameworks and flows presented in this chapter.

For these reasons, this section shows how SystemC-AMS can be extended to support behavioral and conservative descriptions. Instead of adding a new abstraction level (with corresponding libraries and classes), the adopted approach uses SystemC-AMS existing primitives in a novel way [28]. Note that, due to the limitations of SystemC-AMS, supported models are strictly linear and time-invariant.

The starting point of the methodology is a Verilog-AMS behavioral description. In Verilog-AMS, a circuit is modeled as an abstract graph of nodes connected by branches [24]. System state is defined in terms of voltages (`V()`) and currents (`I()`) associated with nodes and branches. Relationships between nodes are modeled with algebraic and differential equations, called *simultaneous statements*.

Since SystemC-AMS is less expressive than Verilog-AMS, any Verilog-AMS simultaneous statement is reproduced by connecting a number of ELN elements. Given a Verilog-AMS description, each simultaneous statement is divided into basic contributions by finding the largest sub-equation that can be represented by a single ELN object. In linear and time-invariant descriptions, this corresponds to breaking the equation into the single addends.

Each addend is then mapped to the most suitable ELN primitive. As an example, an instance of the `sca_vsource` primitive is used to reproduce independent voltage sources, e.g., `V(a) <+ +8.01`. On the other hand, an instance of the `sca_vccs` primitive reproduces voltage controlled current sources, e.g., `I(a) <+ +4.02 V(b)`. ELN primitives must then be connected to reproduce the relationship expressed by the starting simultaneous statement. If the term on the left-hand side of the simultaneous statement is a current, SystemC-AMS instances are connected in parallel. Else, if the term is a voltage, instances are connected in series, by adding intermediate components. Figure 3.12 exemplifies these concepts on a simultaneous statement including a voltage controlled current source, a current controlled current source, and an independent current source.

Differential contributions require a more complex approach, as they model a derivative (or integrative) relationship between the current or voltage of two separate circuit nodes. SystemC-AMS, on the other hand, restricts differential behaviors to dependencies on single network nodes, through the adoption of capacitors (`sca_c` ELN module) or inductors (`sca_l`). To overcome this limitation, it is necessary to introduce an intermediate node that has no physical correspondence in the circuit, but that is rather used for describing the differential dependence. The node is connected to an inductor in case of a derivative construct (e.g., `I(a) <+ ddt(+4.02 V(b))`) and to a capacitor in case of an integrative construct (e.g., `I(a) <+ idt(+4.02 V(b))`). Suitable ELN primitives are then used to bind the evolution of the intermediate node to the nodes involved in the starting differential contribution.

As the application of the proposed approach may be tedious and error-prone, and thus prevent the application to industrial-size case studies, the whole methodology has been automated on top of the HIFSuite framework.

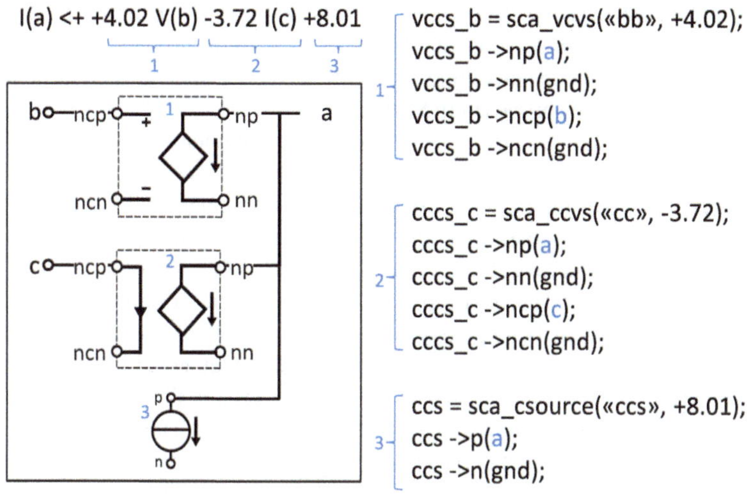


**Fig. 3.12** Example of mapping of a Verilog-AMS simultaneous statement to SystemC-AMS. The simultaneous statement includes a voltage controlled current source (term 1, mapped to an instance of `sca_vccs`), a current controlled current source (term 2, mapped to a `sca_cccs`), and an independent current source (term 3, mapped to a `sca_csource`). Since the *left-hand side* of the simultaneous statement is a current construct, all ELN instances are connected in parallel. Non-connected terminals are connected to ground

Integration Strategies and Challenges

The code generation solutions presented in this section tackle the heterogeneity of smart systems by adopting a common language (i.e., SystemC and its extension), still preserving the heterogeneity in terms of MoC. However, interaction between different MoCs does not rely on manual, error-prone synchronization approaches, as for the functional level (Sect. 3.4.1). All synchronization is indeed transferred to the simulation kernel, that satisfies requests from all MoCs and abstraction levels.

Synchronization correctness is thus guaranteed by the underlying SystemC simulation kernel, that natively masters heterogeneous requests and takes care of synchronization issues between its extensions and MoCs. Furthermore, native converters allow to perform data conversion and to propagate events from one MoC to the other, without any manual intervention from the user. Still, the heterogeneity in terms of MoCs affects simulation performance, as data and synchronization conversion imply a computation overhead. Thus, the simplicity of integration comes at a price of simulation performance.

#### 3.4.2.2 SystemVue-Based Simulation

SystemVue is an environment designed for easing the integration process. Its execution semantics is based on the synchronous dataflow MoC. As such, system behaviors are described by interconnecting basic blocks, expressing a functionality. The strength of SystemVue is that it provides predefined blocks as well as a C++

API to create libraries of custom components that can be included in a system simulation together with components shipped with SystemVue. This allows to easily integrate any C++ code, including manually designed code and code generated with the methodologies proposed in Sect. 3.4.1.

The first step to integrate a C++ external component in SystemVue is to specify its interface as names and data types of all the inputs, outputs, and parameters. The *interface of a SystemVue node* implemented in C++ is composed by a set of variables that are then specified to belong to the interface using the macros: `DEFINE_MODEL_INTERFACE`, `ADD_MODEL_OUTPUT`, and `ADD_MODEL_INPUT`. The data types of these variables, in order to be accepted by the macros, have to belong to a well-defined subset of the available C data types. Some data types, such as circular buffers, are implemented in the SystemVue support library. The other available data types are a subset of the C/C++ data types, that does not include the standard `unsigned integers`. This can be an issue, as normal `unsigned int` data types do not ensure that the span of data representation is the same on different architectures. For this reason, in order to assure the predictability of the number of bits used to represent data on the interface, every variable is declared as `double`. Then, before any computation step, the data read from the interface is assigned to a data structure using standard `Integer` and `Boolean` variables for computation. After the computation, the variables of the data structure are copied into the output variables. Figure 3.13 gives a sketch of the C++ code generated by HIFSuite for SystemVue. The left-hand side of the figure focuses on

```
class component : public
    SystemVueModelBuilder::DFModel
{
   unsigned int cycles;
   ...
   double input_1;
   double input_2;
   double output;
   ...
   struct component_iostruct{
      uint32_t input_1;
      bool input_2;
      uint32_t output;
   } io_exchange;
   ...
   DEFINE_MODEL_INTERFACE(component)
   {
      ADD_MODEL_INPUT(input_1);
      ADD_MODEL_INPUT(input_2);
      ADD_MODEL_OUTPUT(output);
      return true;
   }
   ...
```

```
   ...
   bool Run(){
      component_iostruct io_exchange(
         0UL,
         false,
         0UL
      );
      io_exchange.input_1 = input_1;
      io_exchange.input_2 = input_2;
      simulate(&io_exchange, cycles);
      output = io_exchange.output;
      return true;
   }
   ...
   void simulate(
      component_iostruct * io,
      unsigned int & cycles )
   {
      // A2tool generated implementation
      ...
   }
   ...
};
```

**Fig. 3.13** Overview of the SystemVue-compliant C++ generated by HIFSuite

the interface and it shows the declaration of the interface variables, the input/output data structure, and the interface declaration. On the right-hand side of the figure the `Run` method exemplifies the usage of input/output variables and data structure.

In SystemVue, functionality is implemented in terms of four functions:

- `Setup()` is used to specify the rate of each port, in particular when using circular buffers, in the node interface. The default value is uni-rate, and it is not mandatory to implement this function.
- `Initialize()` is executed during the initialization of the dataflow, thus should be used to run all the initialization code necessary to the node functionality.
- `Run()` is the main method, as it contains the functionality that has to be executed at every simulation step. Its execution is scheduled by SystemVue, according to the dataflow structure, and the rate of the input/output ports of the node.
- `Finalize()` performs any post-simulation coding that the model needs to perform, such as closing file or de-allocating memory.

In order to respect this interface, C++ code generation techniques must be customized and extended to ensure SystemVue support. As an example, the code generated by HIFSuite uses the `Initialize()` method to reset all variables and data structures of the component. The `Run()` function, as depicted on the left part of Fig. 3.13, handles the input/output as discussed above and it calls the code generated by $A^2T$ (i.e., simulate) to emulate component evolution, passing the input/output structure as parameter. When the simulate function returns, the output variables are written according to computed component evolution.

A final integration issue arises whenever components adopt different MoCs. In SystemVue, synchronization and communication among different nodes is based in SDF, that forces the insertion of a delay in every loop among different components. Thus, it is necessary to insert delays to break the loops between connected components, for instance, between a bus and a CPU or between bus and peripherals. However, the generalized insertion of such delays can produce synchronization problems due to the modification of simulation delays that usually guarantee the correct behavior of a digital system. For this reason, digital components in loop are automatically merged by HIFSuite in a single component and abstracted with $A^2T$ as a single component.

By following these guidelines, SystemVue easies the integration of existing code, as the designer must simply match the APIs for the designed components, while the synergy with HIFSuite automatically translates pre-design digital and analogue components and all synchronization issues are left to the simulation kernel.

## 3.5 Experimental Validation of Proposed Methodologies

The goal of this section is to support the proposed analysis and methodologies with experimental evidence. To this extent, the proposed examples focus on single code generation techniques and on the simulation of a complex smart system case study achieved through SystemVue.

**Table 3.1** Abstraction alternatives of digital components for functional and transactional simulation

| Design | Modelsim (VHDL/ Verilog) T (s) | SystemC RTL, SystemC data types T (s) | Abstract C++, HDTLib data types types (SystemC top) T (s) | S (x) | Abstract C++, C++ native data (SystemC top) T (s) | S (x) | Abstract C++, C++ native data types (pure C++ top) T (s) | S (x) |
|---|---|---|---|---|---|---|---|---|
| AES | 72.3 | 850.9 | 332.5 | 2.6 | 8.0 | 106.4 | 7.1 | 119.8 |
| Camellia | 1823.7 | 25,433.3 | 9022.6 | 2.8 | 8.0 | 3179.2 | 3.3 | 7707.1 |
| DES56 | 707.5 | 7608.5 | 1941.1 | 3.9 | 8.5 | 895.1 | 4.6 | 1654.0 |
| SHA512 | 1758.9 | 6302.1 | 2452.4 | 2.5 | 12.6 | 371.2 | 3.4 | 1377.2 |
| XTEA | 171.8 | 975.2 | 260.9 | 3.7 | 18.0 | 54.2 | 3.4 | 286.8 |

### 3.5.1 *Validation of HIFSuite-Based Language Conversion Techniques*

The automatic abstraction of digital components to SystemC/TLM and to C++ plays a key role in the simulation of a smart platform at both the functional and the transactional levels. Thus, its effectiveness must be evaluated in depth.

Table 3.1 reports simulation time (`T(s)`) for some VHDL and Verilog digital components, together with the speedup achieved through the automatic abstraction by A$^2$T with the support of HDTLib or DDT for data type abstraction. The reference simulation time is generated by Modelsim (column *Modelsim*). The generated code may be managed through either a SystemC top-level module (columns labeled with *SystemC top*) or a C++ main simulation file (*pure C++ top*). This distinction allows to analyze all the scenarios outlined in Fig. 3.5, thus covering both the functional abstraction level (single/multiple MoC) and the transactional abstraction level (through the adoption of SystemC or SystemVue for component aggregation).

Results clearly conclude that the automatic abstraction of digital components is extremely efficient (up to three orders of magnitude in speedup) in the case of RTL modules converted to C++ for single MoC functional simulation or for SystemVue-based transactional simulation. In the other cases, the effectiveness of the abstraction process is limited on single components, but it still produces a simulation advantage whenever the platform model must be built by aggregating different components.

### 3.5.2 *Validation of the Mapping of Analogue Conservative Descriptions to SystemC-AMS*

Mapping of analogue conservative descriptions to SystemC-AMS proved to be a complex step, due to the requirements in terms of construct coverage and of application of energy conservation laws. In order to prove the effectiveness of the overall methodology, we applied the overall approach to a complex industrial case

**Fig. 3.14** Three-dimensional model of the accelerometer in the MEMS+ design simulator

**Table 3.2** Characteristics of the original Verilog-AMS MEMS design

| Lines of code | | 89 |
|---|---|---|
| Equations | Voltage sources | 10 |
| | Current sources | 15 |
| Node declarations | Interface | 14 |
| | Internal | 14 |
| Contributions | Independent | 4 |
| | Voltage | 59 |
| | Current | 0 |
| | Derivative | 12 |
| | Integrative | 0 |

study, developed in the context of the SMAC project. Application to this industrial case studies was eased though the implementation of an automatic tool, called ABACuS (Analogue BehAvioural Conservative SystemC-AMS), that leverages HIFSuite to ease the conversion process.

The adopted case study is a *two-dimensional MEMS accelerometer* implemented in Verilog-AMS by means of the MEMS design platform MEMS+, that supports automatic Verilog-AMS code generation [12], starting from three-dimensional physical models as the one depicted in Fig. 3.14. Table 3.2 reports the main characteristics of the MEMS design, both in terms of simultaneous statements and of types of contributions. The MEMS design features most of types of supported contributions, thus showing the application and validation of a significant part of the methodology on a single case study.

Table 3.3 shows the results of the application of ABACuS to the MEMS design. The table shows the number of lines of code of the resulting SystemC-AMS implementation, the number of added nodes and of instances of SystemC-AMS primitives. The number of lines of codes is increased tenfold (precisely, 11.12x),

**Table 3.3** Characteristics of the generated SystemC-AMS MEMS design

| Lines of code | | 1474 |
|---|---|---|
| Added node declarations | | 12 |
| SystemC-AMS primitive instantiations | `sca_r` | 93 |
| | `sca_vsource` | 4 |
| | `sca_vcvs` | 32 |
| | `sca_ccvs` | 0 |
| | `sca_csource` | 0 |
| | `sca_vccs` | 48 |
| | `sca_cccs` | 0 |
| | `sca_l` | 12 |
| | `sca_c` | 0 |

**Table 3.4** Characteristics of the execution of ABACuS on the MEMS design

| Overall | | 17.48 s |
|---|---|---|
| HIFSuite tools | Conversion to HIF | 1.86 s |
| | Conversion to SystemC-AMS | 7.81 s |
| ABACuS | Node management | 0.94 s |
| | Division into contributions | 0.29 s |
| | ELN component instantiations | 6.58 s |

as the SystemC-AMS generated by the methodology is more verbose than Verilog-AMS. Each contribution requires the instantiation of the ELN primitive, plus the corresponding explicit port binding. Furthermore, the number of ELN primitives is higher than the number of Verilog-AMS contributions. This is due to the presence of 12 derivative contributions in the original Verilog-AMS code. Each such contribution determines the instantiation of three ELN primitives (as explained in Sect. 3.4.2.1). As a result, of the 188 resulting SystemC-AMS ELN instances:

- 93 correspond to resistors added to connect each SystemC-AMS node to ground;
- 59 correspond to voltage source contributions;
- 36 are generated by the 12 derivative constructs, that determine also the declaration of 12 additional internal nodes.

Fast code generation is a major advantage of the proposed approach. Table 3.4 highlights that code generation is almost instantaneous (17.48 s overall), and that most of the effort is spent in the HIFSuite conversions (55 %). The most costly step of ABACuS lies in the mapping from Verilog-AMS contributions to ELN primitives and in their instantiation (37 %). On the other hand, node management and the separation of Verilog-AMS equations into single contributions are almost immediate.

The generated code was validated by comparing its execution *w.r.t.* the original Verilog-AMS code, run by using the Questa simulator [23]. SystemC-AMS simulation was run by adopting the same input stimula of the Verilog-AMS implementation, and with a 1us timestep. SystemC-AMS proved to be slightly faster than the Verilog-AMS execution (28.02 s and 33.72 s, respectively). At the same

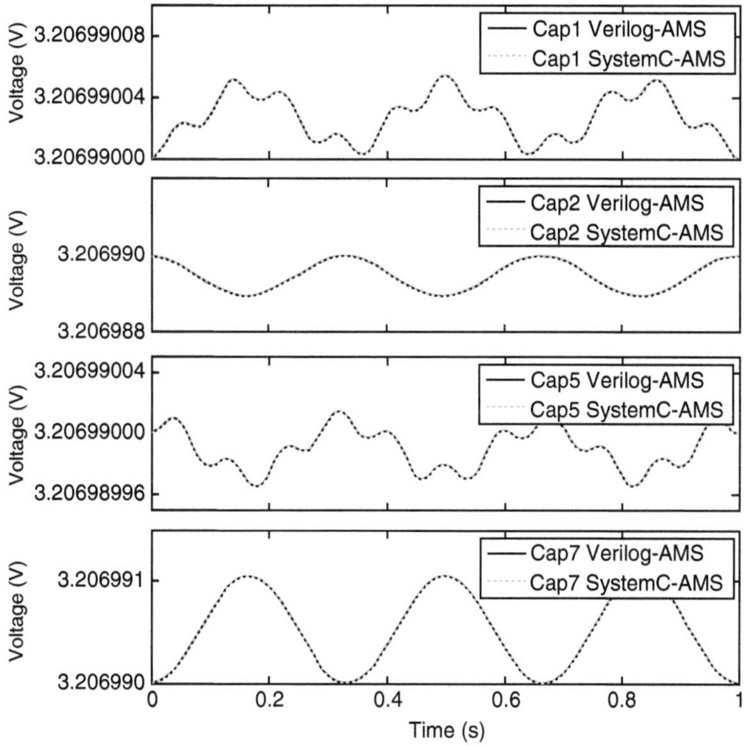


**Fig. 3.15** Evolution of the MEMS outputs for Verilog-AMS (*solid*) and SystemC-AMS (*dashed*)

time, the average error in the computation of the MEMS outputs is 0.02 %. This confirms the visual accuracy evident from Fig. 3.15, where the Verilog-AMS and SystemC-AMS curves are almost totally overlapping. The small error is due to the different management of time in the two simulators: SystemC-AMS adopts a fixed timestep, while Verilog-AMS can adapt the length of the timestep over time, thus reaching a higher accuracy. The low error rate highlights the effectiveness of the generated code, both in terms of accuracy and of simulation speed.

### *3.5.3 Adoption of SystemVue for a Heterogeneous Case Study*

The final example collects all previous results to show a transactional level simulation of a smart system based on SystemVue integrating a number of heterogeneous components. The starting point is complex heterogeneous smart system, developed with the goal of representing a generic smart system. The system, called *open source test case* (OSTC), includes eight modules covering digital HW, embedded SW, RF-transceiver, network elements, and a MEMS sensor (i.e., the accelerometer).

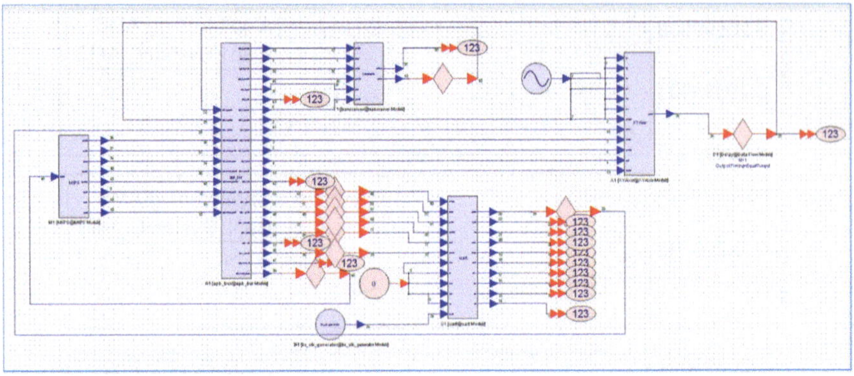

**Fig. 3.16** SystemVue schematic of the OSTC. The left-most component represents the sub-system memory and CPU. This is connected to the component implementing the bus. Components on the right-hand side implement the peripherals. Red rhombuses are delays introduced to break dataflow loops. Red circles represent sinks collecting the outputs of the OSTC

**Table 3.5** Simulation time for the three different simulation scenarios in SystemVue

| Scenario | Simulation time (s) | Speed-up |
|---|---|---|
| Co-simulation of all digital HW | 278.59 | – |
| Co-simulation of one digital HW | 153.23 | 1.8× |
| C++-based simulation | 36.32 | 7.7× |

Such modules are extremely heterogenous in terms of language, as they are described in SystemC, VHDL, Verilog, Verilog-AMS and C++. An exhaustive description of the OSTC will be the focus.

Figure 3.16 shows the SystemVue representation of the OSTC. Each module has been imported in SystemVue after its abstraction to C++, performed by using HIFSuite. SystemVue supports co-simulation, thus allowing the comparison of the following scenarios:

- Co-simulation of all digital HW components;
- Co-simulation of one digital HW component;
- Homogeneous C++-based simulation.

The simulation scenario used for all the models simulates 100 ms of system execution, with a timestep of 100 ns. The inputs of the accelerometer are sinusoidal stimula, and the software application is pre-loaded in the memory. The software takes care of system boot and peripheral initialization. Then, the application repeatedly reads data from the accelerometer, computes the data, and sends the results to the digital hardware and the network interface.

Table 3.5 shows the time needed to simulate the three different scenarios. What appears clear from these results is that the number of simulators instantiated, hence the number of co-simulation interfaces employed, heavily impacts performance. In particular, it worth notice that, in this case, every co-simulation interface (two in the case of the first entry of the table, one in the second), seems to introduce around 120 s overhead *w.r.t.* the simulation without co-simulation interface, thus introducing an overhead of about 80 %. As a result, the impact of interfaces and conversion layers between different tools seems highly relevant and strictly dependent on the number of used interfaces and external tools. The limited speed-up is mainly affected by the low abstraction capability of the two main digital components of the OSTC. Such components are indeed described at gate level rather than at RTL, thus the abstraction to C++ is not extremely effective. Higher speedups can be obtained by using real RTL components, such as the ones reported in Table 3.1.

## 3.6 Concluding Remarks

This chapter provided a formalization of the abstraction levels and design domains of a smart system. This taxonomy allows to identify a precise role in the design flow for co-simulation and simulation scenarios, and to examine the impact of heterogeneous or homogeneous MoCs. Moreover, a methodology has been proposed to move from the co-simulated heterogeneity to a simulatable homogeneous representation of the entire smart system at two level of abstraction: functional level and transactional level. At functional level, all components are implemented in C++, with the goal of understanding the role of the underlying synchronization and simulation semantics and their overhead on simulation performance. At transactional level, two widespread simulation frameworks, i.e., SystemC and SystemVue, have been adopted to ease code integration, even in presence of very heterogeneous design flows.

## References

1. Accellera Systems Initiative, SystemC (2015), accellera.org/downloads/standards/systemc
2. Accellera Systems Initiative, SystemC-AMS 2.0 Standard (2015), accellera.org/downloads/standards/systemc
3. Accellera Systems Initiative, SystemC TLM (Transaction-Level Modeling) (2015), accellera.org/activities/working- groups/systemc-tlm
4. Agilent Technologies, SystemVue Electronic System-Level (ESL) Design Software (2015), www.home.agilent.com/en/pc-1297131/systemvue
5. F. Balarin, Y. Watanabe, H. Hsieh, L. Lavagno, C. Passarone, A. Sangiovanni-Vincentelli, Metropolis: an integrated electronic system design environment. Computer **36**(1), 45–52 (2003)
6. N. Bombieri, G.D. Guglielmo, M. Ferrari et al., HIFSuite: tools for HDL code conversion and manipulation. EURASIP J. Embed. Syst., 1–20 (2010)
7. N. Bombieri, F. Fummi, G. Pravadelli, Abstraction of RTL IPs into embedded software, in *ACM/IEEE Design Automation Conference* (2010), pp. 24–29

8. N. Bombieri, F. Fummi, G. Pravadelli, Automatic abstraction of RTL IPs into equivalent TLM descriptions. IEEE Trans. Comput. **60**(12), 1730–1743 (2011)
9. N. Bombieri, F. Fummi, V. Guarnieri, F. Stefanni, S. Vinco, HDTLib: an efficient implementation of SystemC data types for fast simulation at different abstraction levels. Des. Autom. Embed. Syst. **16**(2), 115–135 (2012)
10. F. Bouchhima, M. Briere, G. Nicolescu, M. Abid, E. Aboulhamid, A SystemC/Simulink co-simulation framework for continuous/discrete-events simulation, in *Proceedings of the IEEE International Behavioral Modeling and Simulation Conference* (2007), pp. 1–6
11. J.C. Butcher, *Numerical Methods for Ordinary Differential Equation* (Wiley, New York, 2003)
12. Coventor, Inc., MEMS+: MEMS Simulation Software (2015), www.coventor.com/mems-solutions/products/mems
13. G. De Micheli, R. Ernst, W. Wolf, *Readings in Hardware/Software Co-Design* (Morgan Kaufmann, San Francisco, 2001)
14. L. Di Guglielmo, F. Fummi, G. Pravadelli, F. Stefanni, S. Vinco, A formal support for homogeneous simulation of heterogeneous embedded systems, in *Proceedings of the IEEE International Symposium on Industrial Embedded Systems* (2012), pp. 211–219
15. EDALab s.r.l., HIFSuite (2015), www.hifsuite.com
16. G. Frehse, PHAVer: algorithmic verification of hybrid systems past HyTech, in *Hybrid Systems: Computation and Control*. Lecture Notes in Computer Science, vol. 3414 (Springer, Berlin/Heidelberg, 2005), pp. 258–273
17. F. Fummi, M. Loghi, M. Poncino, G. Pravadelli, A cosimulation methodology for HW/SW validation and performance estimation. ACM Trans. Des. Autom. Electron. Syst. **14**, 23:1–23:32 (2009)
18. D.D. Gajski, N.D. Dutt, A.C.-H. Wu, S.Y.-L. Lin, *High-Level Synthesis: Introduction to Chip and System Design* (Kluwer Academic Publishers, Norwell, 1992)
19. L.D. Guglielmo, F. Fummi, G. Pravadelli, F. Stefanni, S. Vinco, UNIVERCM: the UNIversal VERsatile computational model for heterogeneous system integration. IEEE Trans. Comput. **62**(2), 225–241 (2013)
20. T. Henzinger, The theory of hybrid automata, in *IEEE Symposium on Logic in Computer Science (LICS)* (1996), pp. 278–292
21. C.-J. Hsu, J. Pino, F.-J. Hu, A mixed-mode vector-based dataflow approach for modeling and simulating LTE physical layer, in *Proceedings of the ACM/IEEE Design Automation Conference* (2008), pp. 18–23
22. E.A. Lee, Overview of the Ptolemy project (2001), ptolemy.eecs.berkeley.edu
23. Mentor Graphics, Questa Advanced Simulator (2015), www.mentor.com/products/fv/questa
24. S. Mijalkovic, Advanced circuit and device modeling with Verilog-A, in *Proceeding of the IEEE Microelectronics* (2006), pp. 439–442
25. P. Schneider, C. Bayer, K. Einwich, A. Kohler, System level simulation - a core method for efficient design of MEMS and mechatronic systems, in *Proceedings of the IEEE Systems, Signals and Devices* (2012), pp. 1–6
26. The MathWorks Inc., Stateflow: Design and Simulate State Machines and Control Logic (2007), www.mathworks.com/products/stateflow/
27. B. van Beek, The Compositional Interchange Format: Introduction (2015), se.wtb.tue.nl/sewiki/media/vanbeek/cifintro.pdf
28. F.F.S. Vinco, M. Lora, Conservative behavioural modelling in SystemC-AMS, in *IEEE/ECSI FDL* (2015), pp. 1–8

# Chapter 4
# Energy-Efficient Digital Processing via Approximate Computing

**Daniele Jahier Pagliari, Massimo Poncino, and Enrico Macii**

## 4.1 Introduction

While the most characteristic feature of a smart system is its capability of *sensing* a set of environmental quantities and *actuating* appropriate actions in response to those signals, it is obvious that a significant part of its functional operations is involved with the *elaboration* of the information carried by the signals [14]. This elaboration is usually done after converting the analog, asynchronous environmental signals into the digital domain.

Part of the smartness of a smart system is therefore expressed in the autonomous and transparent operation based on closed loop control and predictive capabilities, as well as improved signal processing technologies. The former functions are normally carried out by a micro-controller or processor core, whereas the latter ones rely on either a digital signal processor (DSP) or an application-specific integrated circuit (ASIC). Hybrid architectures, that combine one or more general purpose CPUs with one or more hardware *accelerators* are also increasingly popular [11].

Such "processing" dimension, coupled with the energy-autonomous nature of these systems put significant emphasis on their energy efficiency [4, 12]. Measures for reducing energy (and power) consumption vary according to the engineering domain of the component being considered. In the computing subsystem, classical low-power techniques for processors and digital circuits can be fruitfully exploited [35]. In this chapter, however, we focus on the explicit signal processing task and show how we can effectively leverage an emerging design paradigm called approximate computing [20, 52].

D. Jahier Pagliari (✉) • M. Poncino • E. Macii
Politecnico di Torino, Corso Duca degli Abruzzi, 24, 10129 Torino, Italy
e-mail: daniele.jahier@polito.it; massimo.poncino@polito.it; enrico.macii@polito.it
www.polito.it

 
N. Bombieri et al. (eds.), *Smart Systems Integration and Simulation*,
DOI 10.1007/978-3-319-27392-1_4

Approximate computing has its foundations in the tradeoff between quality and energy. It is based on the principle that, accepting a controlled degradation in output quality, energy consumption can be reduced significantly, by changing either the implementation of a computing device or its operating conditions. In smart systems, inputs to the computing subsystem often consist of physical data sampled by some type of sensor. This information is inherently imprecise, both because of environmental noise and measurement errors, and because of the limited precision of transducers and analog-to-digital converters. Therefore, small approximations in computation may be accepted, as their impact on the final output quality may be negligible with respect to the effect of input imprecisions.

Moreover, some of the outputs in smart systems are power actuators, which normally have much longer time constants with respect to electronic elements. As a consequence, an error that manifests only *intermittently* for short amounts of time is automatically "filtered-out" by the actuators.

Both aspects show that the maximum computation accuracy constraint is often overly restrictive in smart systems applications. In other words, these applications are *error resilient*, i.e. can tolerate some computational errors without a significant impact on the quality of results. This property makes them the ideal targets for the approximate computing methodology. In fact, if the design constraints of the processing subsystem are set correctly, approximations can be made negligible with respect to input errors and/or to the resolution of outputs, effectively reducing energy consumption without impacting the output quality in a significant manner.

## 4.2 Error Resilient Computing Paradigms

In modern electronics, energy has become a primary concern, due in particular to the widespread diffusion of mobile, battery-powered devices. Researchers identified *error resilience* as a common characteristic often found in applications performed by such devices, including smart systems, that can be exploited to optimize their energy efficiency [10].

### *4.2.1 Error Resilience*

Error resilience, or tolerance, can be defined as the capability of tolerating some errors without a significant impact on output quality. There are several factors that affect the resilience of an application. In this section, we try to categorize the most recurring ones (see Fig. 4.1).

**Noisy or Redundant Input Data** One cause of resilience is the fact that a system deals with inputs affected either by errors or by some form of environmental noise.

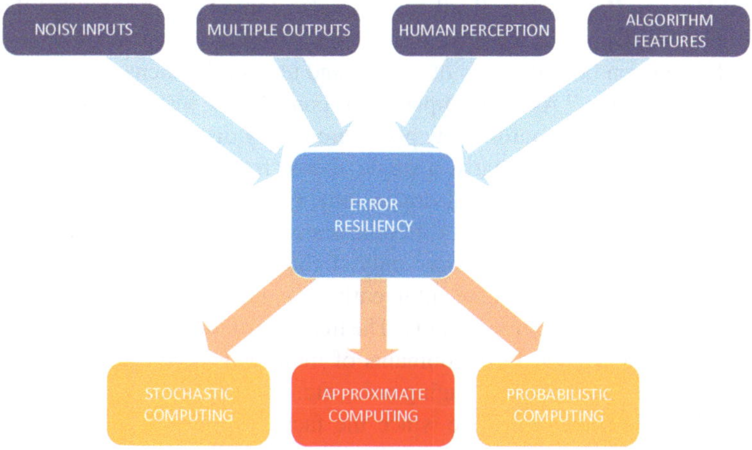


**Fig. 4.1** Overview of error resilience features and error resilient design paradigms

In this situation, errors in computation can be tolerated, as long as they are negligible with respect to the imprecisions on inputs. The latter, in fact, effectively constitute an upper bound on the accuracy that can be obtained at the outputs. One example of this situation has been already anticipated in Sect. 4.1, and corresponds to applications that process data coming from analog sensors, which are inherently affected by environmental noise. Similarly, computing elements inserted in the decoding chain of a communication system also fall in this category, as their inputs are affected by noise and interferences in the channel. Another family of resilient applications are those that process a redundant input data set, as is typical in several machine learning tasks. In this case, some computations can be approximated or even skipped completely, because they do not add information (quality) to the results.

**Absence of a Unique Golden Output** For many applications, the definition of optimal "golden" results is informal or fuzzy. This may happen because multiple outcomes are equivalently valuable for the purposes of the system mission. Alternatively, the optimality of the results produced by a computation can be inherently unknown, because of the presence of random, semi-random, or heuristic operations. In the latter case, slightly perturbing the outputs of an internal operation might neither worsen nor improve the quality of the final results. The first property is present in many data mining tasks. As an example, two similar outputs produced by a web browser search engine might be considered equally good for the end-user, and it is very difficult to distinguish the optimal. On the other hand, applications in the domains of optimization and operation research often display the second feature. In literature, the broad family of recognition, mining, and synthesis (RMS) applications is often reported as a particularly significant example of error resilience [10].

**Limited Human Perception** Resilience can also come from the fact that the final results of a computing task are often evaluated by people. Human sense organs have a limited resolution both in terms of spacial and temporal dimensions and in terms of discernible "values." As an example, let us consider the visual system. Two full-HD images that differ only for a few pixels are hardly distinguishable to the naked eye (spatial dimension). Similarly, a single altered frame in a video stream is also practically unnoticeable (temporal dimension). Finally, the two images or videos are still perceived identical, even if differences are much more frequent in space and time, as long as the pixels colors are only slightly altered (values dimension). In summary, approximations that are sufficiently *rare*, in space or time, or sufficiently *small* do not affect the perceived quality. The majority of the applications that belong to this category are found in the domains of multimedia and telecommunication, e.g., audio/video compression, imaging tasks, etc. Notice that this feature is partially overlapped with the previous one, since the fact that certain differences are not noticeable by humans can be interpreted as a quality equivalence among multiple outputs.

**Algorithmic Features** Lastly, an application can be resilient because of the inner characteristics of the involved algorithms. In particular, certain computational patterns favor the mitigation or the rejection of errors. A typical example of such patterns is *iterative refinement*, used in many recognition and mining applications as well as for the solution of systems of linear equations [20]. This pattern starts from an inexact initial solution and iteratively improves it. Because of how it is constructed, possible additional error contributions introduced by approximate computations will be reduced as well. Again, there is a partial overlapping between this characteristic and the absence of a unique golden output. For example, pseudo-randomness and heuristic decisions can be also thought of as algorithmic features.

In summary, in a resilient application, the quality of results can be thought of as a *continuous function* of the quality of computations, as opposed to a *boolean* one. Approximate computing is one of the design paradigms that exploits this new dimension for the optimization of computing systems, mostly in terms of energy and power consumption (Fig. 4.1).

### *4.2.2 Error Resilient Paradigms*

In literature, approximate computing is distinguished from other similar approaches, that also leverage error tolerance, such as stochastic and probabilistic computing [20]. In this section we briefly describe the main characteristics and differences among these three perspectives.

#### 4.2.2.1 Stochastic Computing

Stochastic computing (SC) was first theorized in the 1960s by two independent researchers in Europe and in the USA [16, 42]. In this paradigm, information stored in streams of bits is interpreted numerically as the probability of occurrence of the logic value 1. As an example, a 10-bit stream with four 1s and six 0s is interpreted as the rational number 4/10. In formal terms, a string of $n$ bits with $n_1$ bits at logic 1 corresponds to the real number:

$$p = \frac{n_1}{n}. \tag{4.1}$$

It can be seen easily that all numbers in this system belong to the interval [0,1], and that their representation is not unique. For example:

$$(0,0,0,1)_{SC} = (0,1,0,0)_{SC} = (0,0,1,0)_{SC} = 0.25_{10}. \tag{4.2}$$

The initial interest for stochastic computing was motivated by the fact that some arithmetic operations between bit streams can be implemented very efficiently in hardware [2]. Moreover, stochastic streams also have good tolerance properties in response to soft (i.e., transient) faults. In fact, if any of the bits of an $n$-bit string changes value because of a transient fault, the error in terms of the corresponding real number is always $1/n$. On the contrary, in binary the impact doubles for each bit going from the least significant bit (LSB) to the most significant bit (MSB).

However, stochastic computing also has some major drawbacks. First of all, increasing the precision of a calculation requires an exponential increase in the length of bit streams. This affects both the speed of calculations and the bandwidth required for memory access and communication. Moreover, not all possible representations of a real number yield the most accurate result when used as input of an operation. More specifically, *correlation* between the different inputs of an operation generates errors at the outputs. To cope with this issue, random or pseudo-random architectures (e.g., linear feedback shift registers) are used to generate the input stochastic streams in the "most uncorrelated" way possible. The problematic aspect is that the probability of correlation errors increases with the number of levels of logic, and in circuits with feedback. Furthermore, some topological aspects of circuits, such as *reconvergent fanouts*, enforce a strong correlation among streams.

The issues related with data correlation and the large bandwidth requirements constrained the application of SC away from the field of general purpose computing. However, this paradigm was used successfully in several domains, such as neural networks, control systems, and image processing [2]. Recently, SC has regained interest due to its application to the decoding of low-density parity check (LDPC) codes. Many of the most efficient algorithms for this task are probabilistic, and therefore error resilient in nature. It has been shown that fast and low cost implementations of such algorithms benefit from an SC representation of data [18].

#### 4.2.2.2 Probabilistic Computing

The trend labeled probabilistic computing (PC) originates from the increasing reliability issues of integrated circuits as technology scales [41]. Because of physical phenomena such as thermal noise and aging, standard CMOS devices are increasingly unreliable from one generation to another. Therefore, hardware designers and manufacturers put great effort in building *reliable* circuits from *unreliable* physical switches via fault tolerant architectures, thermal aware design, etc.

Probabilistic computing reverses this idea, accepting to work with *unreliable* circuits. To this end, a probabilistic model is formulated, in which each switching device is associated with a probability of correctness $p < 1$, and computing architectures are built on top of this model. The rationale that motivates PC is that accepting a small drop in $p$ can provide a large reduction in energy consumption, hence enabling very efficient computation.

Physical devices exhibiting this type of accuracy vs. energy tradeoff have been first theorized and then manufactured, giving birth to the logic family known as probabilistic CMOS (PCMOS). In parallel, probabilistic Boolean logic (PBL) has been adopted as an abstract model to support designs based on PCMOS [6]. Probabilistic computing devices have been used to design architectures for various error resilient applications, including decision systems, pattern recognition, and cryptography, with promising results in terms of power efficiency [41].

#### 4.2.2.3 Approximate Computing

Approximate computing (AC), which is the main focus of this chapter, is allegedly the most successful error resilient design paradigm to this day [20]. It differentiates from probabilistic computing because the traditional deterministic logic model for switching devices is maintained; AC methods do not assume any probabilistic nature in the underlying physical devices that constitute the computing system. Similarly, AC is also a separate body of work with respect to stochastic computing. In fact, in most instances, information is processed and transmitted in standard binary representations (unsigned, two's complement, etc.).

Approximate computing focuses on building imprecise or inexact circuits with *deterministic* architectures (e.g., based on standard CMOS transistors). Statistical considerations, for example, on input data, are extensively used, but in general, a reliable behavior of the hardware at the physical level is assumed. Hence, it is sometimes referred to also as Inexact Computing [41]. The main goal of AC is, as mentioned, power and energy reduction, but there are also techniques that exploit approximation to achieve performance increase and silicon area optimization.

The success of this approach is partly due to the fact that it relies on existing models and design techniques, modified appropriately to target the error resilient application domain, rather than completely changing the perspective. This makes it more accessible to hardware and software engineers accustomed to standard design flows. As an example, many embodiments of approximate computing are inspired by techniques from the hardware and software reliability literature [7, 22, 46], while others make use of modified versions of standard electronic design automation (EDA) algorithms [40, 47, 50].

In an attempt to build a topology of approximate computing techniques, a first order classification can be done to split them in two groups:

- The first family includes techniques that modify the *architecture* of an existing hardware or software design, introducing approximations in the computation, in such a way that the impact of these modifications on the output quality is statistically negligible, while the power and energy consumption are significantly reduced (e.g., [34]).
- The second group includes techniques in which power reduction is achieved by modifying the *operating conditions* of the target design rather than its structure. Structural modifications can still be present, but in this case they are a way to mitigate the quality loss rather than to directly reduce power (e.g., [22]).

A second classification of approximate computing literature can be done according to the level of abstraction. This is the organization that will be followed in the rest of this chapter to describe the details of some of the most successful techniques. In particular, we can identify: (1) transistor/gate-level hardware techniques, (2) algorithm/architecture-level hardware techniques, (3) processor-level hardware techniques, (4) cross-level hardware techniques, and (5) software techniques. In this discussion, we will mostly focus on hardware approaches, since they are currently the most diffused and effective ones. Software AC will be briefly described in Sect. 4.4.

## 4.3 Error/Quality Metrics

Before describing the details of the different approximate computing techniques, it is important to briefly discuss the matter of quality evaluation. AC is essentially a tradeoff between quality and some other metric, usually power or energy. To seek for optimality, both axes of the design space must be quantifiable. Well-established models are available in literature for the power and energy consumption of digital circuits [35]. In this section, we focus on metrics to evaluate quality, and in particular to those suitable for hardware systems. Since quality is a function of the errors introduced by approximate computations, quality and error metrics are often used equivalently. In general, a quality metric can be obtained by taking the inverse of the corresponding error metric.

### 4.3.1 Error Rate and Error Significance

Quality is mostly a domain-specific metric, and there is certainly not a universal function valid for the optimization of any approximate computing target. However, the overall quality of a computing system is often the result of two concurring factors, which are commonly referred to as Error Rate and Error Significance (or Magnitude) [20].

Error rate (ER) is the *rate* or *frequency* of occurrence of errors. It is most commonly measured as the number of erroneous outputs over the number of total outputs considered, both erroneous ($n_e$) and correct ($n_c$):

$$\mathrm{ER} = \frac{n_e}{(n_e + n_c)}. \tag{4.3}$$

Informally, ER can be considered as the *average probability* of occurrence of an error over the time interval considered for its evaluation [10].

Error significance (ES) defines instead the *severity* of a single error. Since the notion of severity is strictly related to the mission of a system, the number of ES functions found in literature is much larger than for ER. In this context, we mention some of the most common ones. The simplest ES metric is the absolute value of the numerical difference ($D$) between the correct ($v_c$) and erroneous ($v_e$) values of a given output at a given time instant [25, 47]:

$$\|D\| = \|v_c - v_e\|. \tag{4.4}$$

A commonly found alternative definition uses the squared difference $D^2$ instead. The Hamming distance (HD), defined as the number of bits that differ between the binary representations of the correct and erroneous values, is often used for strategies targeting arithmetic circuits [25]:

$$\mathrm{HD} = \sum_{\mathrm{bit}(i)=1} (v_c \oplus v_e) \tag{4.5}$$

where $\oplus$ is the bitwise XOR operator. Finally, the *relative* difference (RD) normalizes the error with respect to the correspondent correct value:

$$\mathrm{RD} = \left| \frac{v_c - v_e}{v_c} \right|. \tag{4.6}$$

### 4.3.2 Composite Metrics and the Fail Small or Fail Rare Concept

In most approximate computing literature, some form of composite metric is used that encompasses the information of both ER and ES [27, 32, 48]. The simplest of such metrics is rate-significance (RS), defined as the product between the rate and the maximum significance over a set of vectors [48]:

$$\mathrm{RS} = \mathrm{ER} \cdot \max(\mathrm{ES}). \tag{4.7}$$

ER and ES are in turn defined according to one of Eqs. (4.3) and (4.4)–(4.6). In *RS*, the contributions of ER and ES can also be assigned a different importance by means of appropriate weighting. A slightly more refined metric is the mean error distance (MED), defined as the mean of $\|D\|$ [see Eq. (4.4)] over the considered set of $N$ output vectors [32]:

$$\mathrm{MED} = \frac{1}{N}\sum_{i=1}^{N} \|v_{c,i} - v_{e,i}\|. \tag{4.8}$$

If errors occurring in any time instant only depend on the value of the correct output in the same instant, MED can be equivalently defined as:

$$\mathrm{MED} = \sum_{j=1}^{M} \|v_{c,j} - v_{e,j}\| \cdot \mathbb{P}(v_{c,j}) = \mathbb{E}\left[\|v_{c,j} - v_{e,j}\|\right] \tag{4.9}$$

where $M$ is the number of possible *values* assumed by the correct output and $\mathbb{P}$ and $\mathbb{E}$ stand for probability and expected value, respectively.[1] The normalized error distance (NED) corresponds to the MED normalized to the maximum possible error on a given output [32]:

$$\mathrm{NED} = \frac{\mathrm{MED}}{\max(\|D\|)}. \tag{4.10}$$

With respect to the unnormalized version, this metric is better to compare the impact of errors on buses with different widths or using different data representations. Very similar to MED is the mean squared error (MSE) [27]:

$$\mathrm{MSE} = \frac{1}{N}\sum_{i=1}^{N} (v_{c,i} - v_{e,i})^2. \tag{4.11}$$

And again, under the assumption that errors only depend on "present" output values:

$$\mathrm{MSE} = \mathbb{E}\left[(v_{c,j} - v_{e,j})^2\right]. \tag{4.12}$$

[1]However, in many AC architectures errors are either affected by previous history of inputs and output values, or by external conditions, hence this definition is not valid.

Up to this point, we mentioned mainly general purpose metrics. However, many of the most used metrics in AC are domain-specific. The most popular in digital signal processing and communication applications is the signal-to-noise ratio (SNR), defined as [1, 22, 27, 46]:

$$\text{SNR} = \frac{\sigma_s^2}{\sigma_w^2} \tag{4.13}$$

where $\sigma_s^2$ is the power of the *signal* component, i.e. of the set of error-free values carrying useful information and $\sigma_w^2$ is the power of the *noise* component, i.e. of the combination of all types of disturbances affecting the measured values. Therefore, when SNR is used, errors due to approximate computations are modeled as a *noise source*. One of the advantages of this metric is that it allows to combine approximations with disturbances caused by other factors, such as limited accuracy of analog components, external radiations, and crosstalk in a communication systems, etc. In that case, assuming that computation inexactnesses and other sources of noise are uncorrelated, the definition of SNR becomes [46]:

$$\text{SNR} = \frac{\sigma_s^2}{\sigma_{\text{noise}}^2 + \sigma_{\text{appr}}^2}. \tag{4.14}$$

This quantity is often expressed in decibel (dB). Notice that SNR is strictly related to MSE, as the "noise" due to approximations is defined as:

$$N_{\text{appr}} = v_c - v_e \tag{4.15}$$

and hence its average power $\sigma_{\text{appr}}^2$ is exactly expressed by Eq. (4.11). A further variant is the peak signal-to-noise ratio (PSNR), which considers the ratio between the maximum value assumed by the correct data ($\max(\|v_c\|)$ and the noise variance. Lastly, many literature works use higher level application-specific models of error/quality [10, 37, 40]. Since these metrics are so specific, they are too many to be enumerated here.

In general, because of the nature of most error resilient applications, combined metrics that consider both the rate and the significance of errors are often the most effective ones. For these applications, the relation between computational errors and measured quality is perfectly summarized by the rule of thumb of "Fail Small or Fail Rare" [10]: the quality of outputs produced by an inexact computing system acting over a set of inputs for a given period of time is acceptable if approximations are either small in significance or rare in time. This is an essential guideline to the design of approximate computing systems.

## 4.4 Approximate Computing Techniques

A schematic view of the main approximate computing techniques organized by abstraction level is shown in Fig. 4.2. In this section we provide a brief overview of each group of approaches, aimed at conveying the general principles, and not intended as a complete survey. We focus mostly on solutions suitable for the optimization of DSP tasks commonly found in smart systems, and only mention the remaining ones. However, we reference the most important publications, in case the reader is interested. Solutions are examined with a bottom–up order, because low level techniques often constitute the building blocks of higher level ones.

### *4.4.1 Transistor/Gate-Level Techniques*

In Fig. 4.2 we use the term netlist-level techniques to identify all those approaches that operate on the transistor or gate-level netlist of circuits. Two main trends are identified in literature at this abstraction. Some researchers have developed approximate functional units, building inexact versions of the most common circuits found in digital datapaths. Others have focused on automation aspects, developing tools for the logic synthesis of approximate circuits.

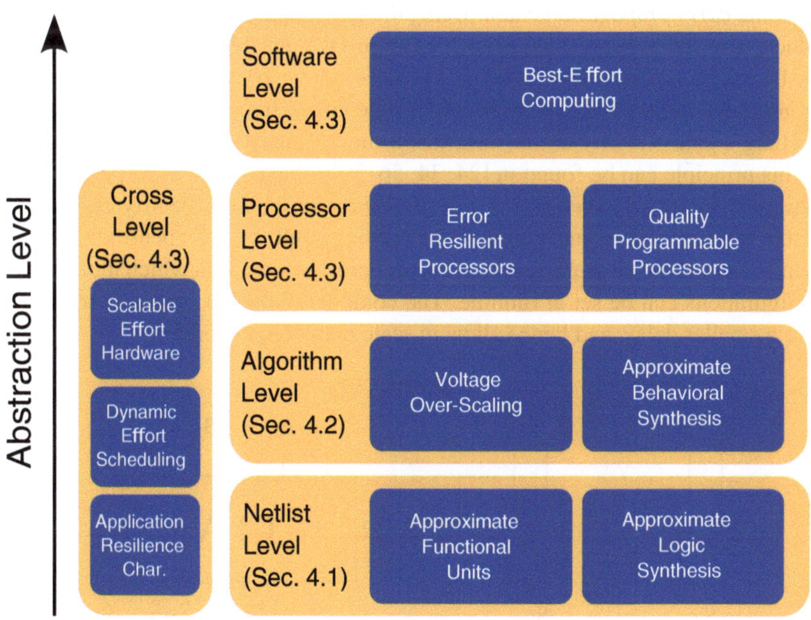


**Fig. 4.2** Overview of the main approximate computing techniques organized by abstraction level

#### 4.4.1.1 Approximate Functional Units

The design of approximate versions of functional units (FU) is one of the most popular applications of AC. Efforts are mainly directed toward the most common datapath elements, namely *adders* and *multipliers*. A comprehensive survey of these techniques can be found in [20].

For what concerns approximate adders, two main categories of approaches can be identified. The first group modifies the structure of a single-bit full-adder (FA) and then generates multiple-bit adders with the modified FA. The second also starts from accurate architectures, but considers them from a coarser point of view, changing the way in which their building elements are connected.

In the category, one of the earliest designs is the so-called *lower-part-OR adder* (LOA) [36], in which some FAs are replaced by a simple OR gate (see Fig. 4.3). In [19, 54] the authors propose a similar idea, but in this case they generate approximate FAs with transistor-level simplifications. These solutions directly reduce area and power (by affecting leakage, internal capacitances, and switching activity) and also decrease the delay of the FA, improving its performance, or alternatively enabling a more aggressive voltage scaling, to further reduce power consumption. Both goals are pursued while minimizing the truth table differences between the approximate versions and the original FA. Approximate FAs are used in the LSBs of a multi-bit adder, while MSBs are summed accurately (e.g., with a carry lookahead adder) to preserve quality, as shown in Fig. 4.3.

The second group of approximate adders is mostly constituted by designs that *break the carry propagation chain*. Being the critical path, breaking it enables delay reduction at the expense of possible errors in the sum output. As mentioned, this can be exploited for voltage scaling. Moreover, the carry chain is also responsible for most switching activity in an adder, due to its frequent glitches. Therefore, reaching signal stability earlier also has a direct impact on power. Adders based on this principle can be found in [24, 34, 56–59]. The authors of [53] use the same idea to design a *variable latency speculative adder*, able to identify errors due to the broken carry chain, and compute the correct result, at the expense of additional latency. In [25] an *accuracy configurable adder* is proposed, in which multiple error configurations can be set at runtime. The configurations that produce larger errors have a smaller delay, and hence allow to scale the voltage more.

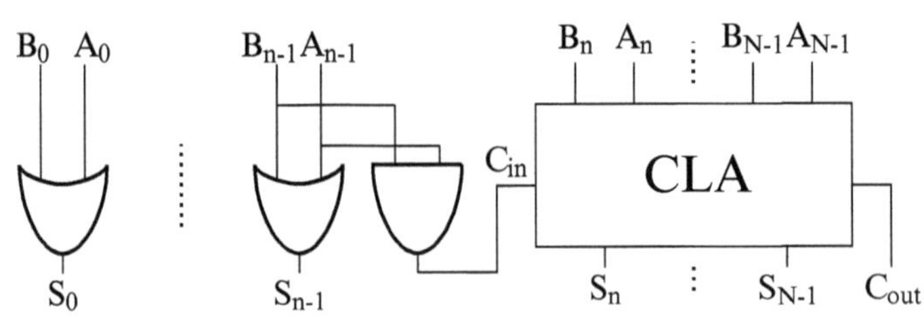


**Fig. 4.3** Lower-part-OR adder block diagram

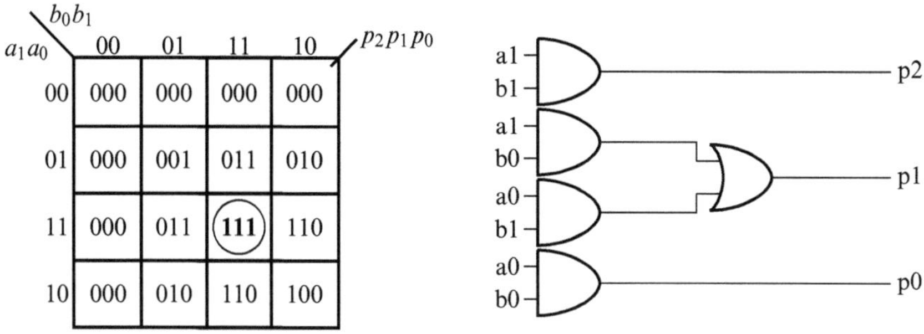


**Fig. 4.4** Modified 2×2 multiplication Karnaugh map and corresponding netlist

Approximate multipliers have a comparatively small literature. Some solutions construct multipliers based on approximate adders [24, 34], while others modify existing array multiplier architectures removing or simplifying the FAs involved in MSBs calculations [30, 36]. A different approach is proposed in [29], where a multi-bit unsigned multiplier is constructed starting from an approximate elementary cell that performs 2-bit multiplication. It is shown that, by changing a single entry in the truth table of the 2×2 multiplier, hardware complexity can be reduced of almost 50 % (see Fig. 4.4). Moreover, the erroneous condition can be easily detected and compensated. Thus, the architecture also supports an accurate mode of operation, used only for quality-critical multiplications, at the expense of additional power. Finally, [33] proposes an advanced design for an accuracy configurable multiplier, based on input pre-processing.

#### 4.4.1.2 Approximate Logic Synthesis

Approximate logic synthesis (ALS) deals with the generalization of the concepts introduced for the approximation of functional units, and applies them to *any* gate-level circuit. The rationale is that automation is indispensable to extend the approach to larger and more complex designs. Some of the first efforts in this direction are found in [47, 48], where methods for ALS are proposed targeting two-level and multi-level circuits, respectively. These solutions tackle the ALS problem directly, using ad-hoc algorithms either to reduce the number of literals [47] or to simplify some nets as if they were redundant [48], based on a given quality constraint.

A fundamental work on ALS is the tool developed in [50], named Systematic methodology for Automatic Logic Synthesis of Approximate circuits (SALSA). SALSA introduces several novelties with respect to previous approaches, including the fact that ALS is transformed into a traditional synthesis problem, allowing to fully leverage the optimization capabilities of commercial tools. Moreover, it

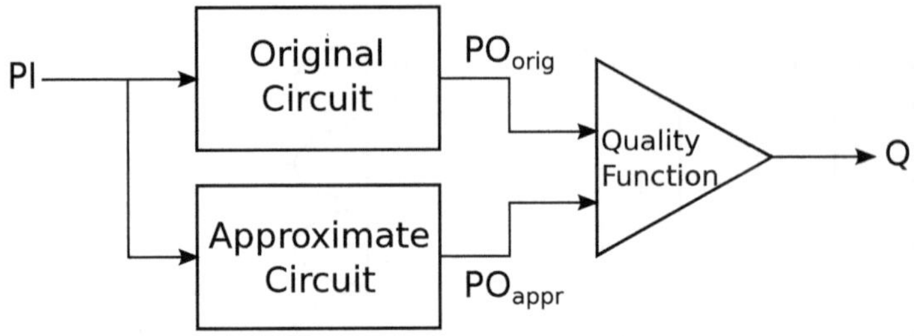


**Fig. 4.5** Quality constraint circuit (QCC) in SALSA

also unlinks the synthesis procedure from a specific quality metric, allowing the designer to set their preferred metric as an input, and making the tool much more flexible. The basic idea of SALSA is to construct the so-called quality constraint circuit (QCC), shown in Fig. 4.5. The triangular block contains a used-defined logic-level implementation of the quality function, and its single-bit output $Q$ is at 1 if and only if the desired quality constraint is respected. SALSA considers the input conditions for which $Q$ is independent from a certain bit in the approximate circuit outputs $PO_{\mathrm{appr},i}$, i.e. the observability don't cares (ODCs) of $Q$ with respect to $PO_{\mathrm{appr},i}$. Since those are the input combinations for which changing the value of $PO_{\mathrm{appr},i}$ does not violate the quality constraint, they are used to simplify the approximate circuit, reducing area, power, and delay. Both the individuation of ODCs and the corresponding simplifications are performed with commercial tools for classic logic synthesis. The process is repeated iteratively on all bits of $PO_{\mathrm{appr}}$, progressively updating the approximate circuit in the QCC. Several optimizations are proposed to reduce the computational complexity of the algorithm. Recently, the SALSA methodology has been extended to consider also sequential circuits [44].

### 4.4.2 Algorithm/Architecture-Level Techniques

Algorithm/architecture-level techniques for approximate hardware can be conceptually defined as those that concurrently optimize a signal processing algorithm and the hardware architecture that implements it. The considered targets can be as simple as a Multiply and ACcumulate (MAC) or as complex as a complete fast Fourier transform (FFT). Therefore, there is not a well-defined distinction between this family of techniques and the previously mentioned netlist-level ones. Architecture-level methods are often the most effective ones for the DSP applications found in smart systems.

#### 4.4.2.1 Voltage Over-Scaling

A large portion of algorithm-level techniques for approximate computing is based on the voltage over scaling (VOS) principle [27]. In a sequential device, at a given clock frequency, the critical voltage $V_{dd,\mathrm{crit}}$ can be defined as the smallest supply voltage that guarantees correct operation, i.e. absence of timing violations in the worst-case operating conditions. VOS-based approaches let the device operate at voltages $V_{dd} < V_{dd,\mathrm{crit}}$, exploiting the fact that timing errors only occur in correspondence of certain combinations of inputs. Therefore, in the majority of cases,[2] the circuit still produces the correct result. The clear advantage is that consumption is reduced significantly, due to the super-linear relation between supply voltage and both dynamic and leakage power [35].

The first embodiment of VOS to produce approximate circuits was originally proposed in [22], and is named algorithmic noise tolerance (ANT). Its main idea is to let an arithmetic or DSP system work in VOS conditions, then leverage an error-free error control (EC) block to detect the occurrence of a timing violation and to *mitigate* its effects on the output quality (see Fig. 4.7). Different types of ANT, in particular for what concerns the implementation of the EC block, are proposed in [23, 45, 46] and others. This technique is one of the most suitable for DSP operations, and is analyzed in detail in Sect. 4.5.

VOS is considered from a different perspective in significance driven design (SDD) [3, 21, 28, 38]. Instead of reducing errors due to VOS, this family of techniques tries to confine their occurrence to *non-significant* computations, i.e. computations that do not impact drastically the output quality. This goal is achieved acting both on the algorithm and on the corresponding hardware architecture. As an example, [3] applies SDD to a hardware implementation of the discrete cosine transform (DCT). Some DCT coefficients are slightly altered (algorithmic modification) in order to improve hardware sharing (architectural modification). This sharing allows to compute high-energy DCT components, which are the most relevant to determine the visual characteristics of the output image, with a reduced delay, while long combinational paths are related to less significant components. Therefore, in VOS conditions, timing errors will only affect the latter.

An important aspect of all VOS designs is that the probability of timing errors must be as small as possible, to limit the impact on quality. These errors are induced by combinations of inputs that activate long combinational paths in the architecture. Therefore, the distribution of path lengths influences significantly the behavior of a system in VOS conditions; the ideal condition is to have few long paths and many short ones, as shown in Fig. 4.6. Two different works consider the problem of easing timing degradation of circuits in VOS [26, 39]. The former proposes a *VOS-friendly* gate sizing algorithm, that changes the path distribution to produce a gradual degradation of quality and enable aggressive over-scaling. In [39]

---

[2]The precise rate of timing errors depends on circuit topology and on the characteristics of input data. Moreover, it can be modified with optimization techniques (e.g., gate sizing).

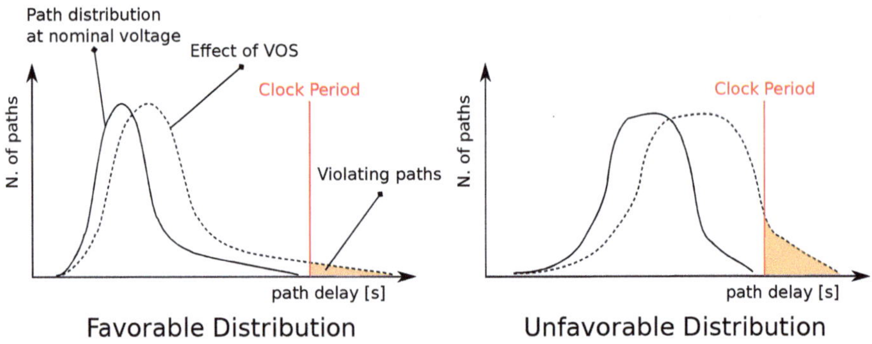


**Fig. 4.6** Impact of slack distribution on VOS-induced errors

the focus is on architectural modifications of some *meta-functions*, i.e. recurring operations common to different error resilient applications, aimed at changing their path distribution to reduce the error rate and significance at a given voltage.

Other literature involves VOS as key feature for power reduction. In [13, 15, 17] the authors proposed techniques and devices (such as the so-called Razor flip-flops) that aim at detecting and avoiding VOS-induced errors. Error avoidance is accomplished by allocating additional clock cycles to let the longest combinational paths stabilize, for instance, via clock gating or pipeline stalling. These techniques are very popular, but since they *avoid* errors rather than accepting them, they do not strictly belong to the approximate computing domain.

#### 4.4.2.2 Approximate Behavioral Synthesis

Recent works have addressed the approximate synthesis of designs described at the behavioral level, i.e. starting from a functional model of the system behavior (typically graph-based) rather than from a net of logic gates. The main effort is documented in [40], where the authors propose a tool named Automated Behavioral Approximate CircUit Synthesis (ABACUS), that produces multiple approximate versions of a circuit starting from an original accurate model in behavioral HDL. ABACUS operates by transforming the HDL in an abstract syntax tree (AST) model, then modifying the AST with a series of *operators*, each of which introduces some approximations while preserving syntactic correctness. Operators include LSBs-truncation, variable to constant substitution, loop unrolling, etc. Moreover, ABACUS can also use models of approximate functional units like those presented in Sect. 4.4.1.1. Approximated ASTs are then transformed back to HDL and processed with a standard synthesis and simulation flow to evaluate their cost (in terms of area and power) and their accuracy. To avoid an exponential number of synthesis and simulation phases, ABACUS does not apply all approximating operands to all feasible locations in the AST. Instead, in each iteration it generates a number of approximated ASTs choosing operands and locations at random, then greedily selects the *best* solution as the starting point for subsequent iterations. Optimality is expressed by a fitness function that weights cost and accuracy components. The entire process is then repeated for a fixed number of iterations. A hard threshold on accuracy is also set, so that solutions that affect quality too much are not synthesized.

### 4.4.3 Other Techniques

In this section we present approximate computing techniques that are less interesting for the context of DSP in smart systems, but nevertheless need to be mentioned to provide a complete view of the state of the art (see Fig. 4.2).

**Processor-Level Techniques** In the domain of programmable processors, AC solutions are mostly based on the acceptance of errors due to process variability, aging, temperature, external radiation, etc. Rather than hardening the entire processor against these errors, only some cores or units are made reliable with fault tolerance mechanisms. Those units manage the control flow and execute critical operations, while error resilient computations are offloaded to unreliable hardware. The total area and power costs related to hardening are thus significantly reduced [31, 55]. A different approach is found in [51], where the authors propose a vector processor in which the "quality-level" of each operation can be set at the software level, thanks to a specific Instruction Set Architecture. Different quality/power levels are obtained in hardware combining reduction of operands bit-widths with power and clock gating.

**Cross-Level Techniques** Another branch of research has focused on methodological aspects related to AC, such as the automatic assessment of error resilience, and the combination (or *synergy*) of techniques at multiple abstraction levels. The first objective has produced the so-called *application resilience characterization* (ARC) framework [10], able to individuate the error resilient computational kernels in an application and evaluate the impact of different AC techniques (e.g., approximate functional units, VOS meta-functions, etc.) on them. The ARC framework is used in [8] to generate *scalable effort hardware* architectures, that combine AC techniques at different abstraction levels to obtain greater power and area savings. A further improvement is found in [9], where a technique called *dynamic effort scheduling* is proposed to set the global quality level at runtime, acting on knobs that affect the different AC techniques involved.

**Software Techniques** Most literature on AC at the software level exploits the previously mentioned *iterative refinement* (also found as *iterative convergence*) computational pattern. A technique called *best effort computing* has been devised for these algorithms in [5, 37], based on principles similar to those of best effort communication in the networking stack [49]. An algorithm is divided into *guaranteed computations*, which are fundamental for the final result, and *optional computations*, which can tolerate errors without a significant impact on output quality. The relaxed correctness requirement on the latter group is exploited in the operating system to perform different types of optimizations, aimed both at performance improvement and cost containment. For example, optional computations can be skipped entirely, to avoid exceeding the power budget, or synchronization between threads can be relaxed to improve performance.

## 4.5 Algorithmic Noise Tolerance for DSP

Many smart systems applications involve some form of digital signal processing (DSP). Depending on the complexity of the task, designers can choose different types of computing devices to implement it. In some cases, a programmable processor is mandatory, while in others the entire application can be implemented with ASIC hardware. Even in the first setting, however, recent trends in industry indicate that computing platforms are becoming increasingly *heterogeneous*, combining general purpose processors with domain-specific hardware *accelerators* [11]. The need for accelerators is driven by the double objective of improving performance while reducing energy consumption; being less flexible, dedicated hardware can be designed to be faster and more efficient than general purpose processors. Moreover, for large sells volumes, heterogeneity also reduces costs, as a given performance level is reached with simpler hardware, thanks to specialization.

In summary, complex smart systems are likely to include some form of specialized hardware module, which is often in charge of the most computationally intensive and energy consuming processing tasks. As mentioned in Sect. 4.1, DSP applications involved in smart systems frequently exhibit error resilience. Consequently, approximate computing configures as a potential source for further energy-driven optimization. Smart systems specialized ASICs are often more complex than a single arithmetic module; examples of frequently found elements include finite impulse response (FIR) and infinite impulse response (IIR) filters, FFT or DCT processors, etc. Multiple of such elements are normally integrated in a single IC. The most suitable approximate computing techniques for this kind of complexity are those working at the algorithm/architecture level (see Sect. 4.4.2), because their high-level view allows to optimize the hardware system in its entirety, rather than each functional unit separately. In particular, *ANT* for the mitigation of errors due to *VOS*is currently the most mature and well-defined approach for approximating computation in specialized hardware modules of this nature [22].

### 4.5.1 *ANT Overview*

The most general scheme of an ANT-based architecture is shown in Fig. 4.7. The gray box corresponds to the hardware that performs the main functionality (e.g., a digital filter) and is referred to as *main DSP* (MDSP) block. Its primary inputs

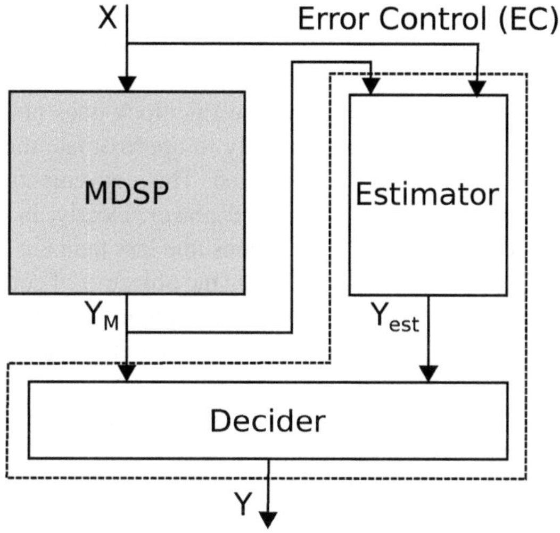


**Fig. 4.7** General scheme of an ANT architecture

and outputs are labeled $X$ and $Y_M$, respectively. The MDSP operates in VOS conditions ($V_{dd} < V_{dd,\mathrm{crit}}$), therefore it consumes significantly less dynamic ($P_{\mathrm{dyn}}$) and leakage ($P_{\mathrm{leak}}$) power than in nominal conditions, according to the well-known power dependencies for CMOS circuits:

$$P_{\mathrm{dyn}} \sim f_{\mathrm{clk}} C_L V_{dd}^2 \tag{4.16}$$

$$P_{\mathrm{leak}} \sim I_{\mathrm{leak}} V_{dd} \tag{4.17}$$

where $f_{\mathrm{clk}}$ is the clock frequency, $C_L$ is the total load capacitance, and $I_{\mathrm{leak}}$ is the leakage current. However, the logic in the MDSP is slowed down by the reduced supply voltage; in fact, the delay of combinational CMOS gates depends on voltage according to the following equation:

$$\tau_d = \frac{V_{dd}}{\beta (V_{dd} - V_{\mathrm{th}})^{\alpha}}, \quad \alpha > 1. \tag{4.18}$$

In ANT, while $V_{dd}$ is reduced below the critical value, the clock frequency ($f_{\mathrm{clk}}$) used to drive sequential memory elements such as flip-flops is not modified, in order to preserve the same performance (latency or throughput) as in nominal conditions. Consequently, the MDSP is subject to input-dependent timing errors; when a long combinational path is activated by a certain input combination, the logic can fail to produce the correct, stable value in time before the following clock edge. To cope with these errors, the MDSP is coupled with an *error control* (EC) block, delimited by the dashed line in Fig. 4.7. The purpose of the EC block is to detect the occurrence of an error due to a timing violation, and limit its effects providing a suitable *approximation* of the correct result at the global outputs of the ANT architecture ($Y$). The combination of MDSP and EC block is sometimes referred to as *soft DSP* [22].

Thanks to the fact that the EC block does not need to *correct* an error, as in a fault tolerant system, but only to approximate the corresponding correct value, its hardware complexity is limited. This conveniently reduces the overheads in terms of additional silicon area and power; clearly, in order to be useful, the complete ANT system in VOS must consume less than the MDSP in nominal supply voltage conditions. Moreover, relaxing the objective of complete error correction also allows to make the EC block timing compliant in VOS. This is a necessary requirement, because it guarantees that the error mitigation performed in the EC block is not affected by errors itself.

The EC block is composed of two main elements, an *estimator* and a *decider*. The estimator is the component that produces an approximation of the correct MDSP output, called $Y_{est}$. In general, the estimation is obtained using information on MDSP inputs and outputs, although the most popular ANT implementations only use either one or the other. The decider, instead, is the module in charge of selecting the output of the MDSP or that of the estimator, depending on the possible occurrence of a timing error. Notice that both MDSP and estimator outputs must be *latched* before being fed to the decider. Otherwise, the decision logic might end up in the critical path, becoming prone to delay errors, and rendering the entire ANT architecture useless. Latching, however, increases of one clock cycle the latency of the system, with respect to the original MDSP.

#### 4.5.1.1 Decision Scheme

ANT is based on two important assumptions on the nature of timing errors due to VOS. Firstly, timing violations must be *rare in time*, so that in most of the cases the MDSP produces the correct output, and that value can be used as global output of the system. This allows to maintain an average output quality which is comparable to that of the MDSP in nominal voltage conditions. The validity of this assumption depends on the selected VOS voltage, on the input data sequence, and on the distribution of timing paths in the MDSP. However, techniques such as slack redistribution can be leveraged to enforce it [26]. Secondly, the errors produced in correspondence of a timing violation must be *large in magnitude*, so that they are easily detectable by the decider. This condition is verified for the large majority of DSP hardware systems. In fact, these systems are composed of a sequence of elementary functional units such as adders and multipliers, for which the critical timing paths are relative to the computation of the MSBs of result. Therefore, when an error occurs, it most likely affects drastically the output.

The fact that errors are large in magnitude allows to select between MDSP and estimator outputs using as discriminating factor the absolute value difference between the two. The estimator is designed to provide a good approximation of the correct output. Therefore, if at a given instant its output is significantly different from that of the MDSP, it is safe to assume that the latter is subject to a VOS-induced error, and vice versa. In summary, the decision scheme is the following:

$$y[n] = \begin{cases} y_M[n], & \text{if } |y_M[n] - y_{\text{est}}[n]| \leq T_h \\ y_{\text{est}}[n], & \text{if } |y_M[n] - y_{\text{est}}[n]| > T_h \end{cases} \tag{4.19}$$

where, $y[n]$, $y_M[n]$, and $y_{\text{est}}[n]$ indicate the values of signals $Y$, $Y_M$, and $Y_{\text{est}}$ at time $n$, respectively.[3] It is important to highlight that this decision scheme effectively mitigates also errors not caused by VOS (e.g., deep sub-micron noise, external radiation, aging, etc.), whose effects on $Y_M$ are significant, as long as the EC block is not affected by the same errors.

#### 4.5.1.2 Quality Metric and Problem Formulation

The main targets of ANT are DSP systems, such as digital filters or FFT accelerators. These systems often process analog data produced by sensors and converted to the digital domain, or received from a communication channel. Both types of input are affected by noise. Therefore, in nominal supply voltage conditions (i.e., in absence of VOS errors) the output of the MDSP can be expressed as:

$$y_{M,0}[n] = s[n] + w[n] \tag{4.20}$$

where $s[n]$ is the signal component, and $w[n]$ is a generic noise component, that may be due to different physical sources. A particularly suitable metric for assessing the processing quality of this type of systems is the SNR, defined in Eq. (4.13). The ANT strategy saves power accepting to introduce errors due to timing violations, which are only partially corrected by the EC block. The residual errors on $Y$ can be modeled as an additional source of noise, as explained in Sect. 4.3. The global output quality of the ANT system can thus be measured (in decibel) with the following expression:

$$\text{SNR} = 10 \log \left( \frac{\sigma_s^2}{\sigma_w^2 + \sigma_{\text{err}}^2} \right) \tag{4.21}$$

where $\sigma_{\text{err}}^2$ (assumed uncorrelated with $\sigma_w^2$) is the power of the additional errors introduced by ANT:

$$\sigma_{\text{err}}^2 = \mathbb{E}[(y_{M,0}[n] - y[n])^2] \tag{4.22}$$

[3] In general, we use capital letters to refer to a signal, and lowercase ones to indicate its value at a given instant in time.

$y[n]$ is defined as in Eq. (4.19), so it takes into account the reduction of VOS-induced errors thanks to the EC block. Clearly, the SNR is not the only possible quality metric for ANT, but given the nature of the considered MDSPs, it is largely the most used. A typical goal is to design the EC block so that errors due to VOS timing violations are negligible with respect to external noise ($\sigma_{err}^2 << \sigma_w^2$).

In general, the problem of designing an optimal ANT architecture for a given MDSP system can be defined as: *finding the VOS voltage and feasible EC block implementation that minimize total power consumption, including EC block overheads, under a minimum quality constraint, and so that the entire ANT system consumes less than the single MDSP in nominal conditions.* In formal terms:

$$\begin{cases} \min[P_{\text{ant}}(V_{\text{vos}}, \mathbf{k})] \\ Q(V_{\text{vos}}, \mathbf{k}) > Q_{\text{des}} \\ P_{\text{ant}}(V_{\text{vos}}, \mathbf{k}) < P_{\text{mdsp}}(V_{\text{nom}}) \end{cases} \tag{4.23}$$

In Eq. (4.23), $Q$ represents a generic quality metric (e.g., SNR) and $Q_{\text{des}}$ is the minimum acceptable quality for the target application. $V_{\text{vos}}$ and $V_{\text{nom}}$ are, respectively, the VOS and nominal (i.e., error free) supply voltages. $P_{\text{mdsp}}$ is the power of the MDSP, and only depends on voltage, since the MDSP hardware is fixed. On the contrary, the power of the entire ANT system ($P_{\text{ant}}$) depends both on the VOS operating point and on the generic vector of parameters $\mathbf{k}$ which determines the implementation of the EC block:

$$P_{\text{ant}}(V_{\text{vos}}, \mathbf{k}) = P_{\text{mdsp}}(V_{\text{vos}}) + P_{ec}(V_{\text{vos}}, \mathbf{k}). \tag{4.24}$$

Depending on how the EC block is designed, $\mathbf{k}$ corresponds to a different set of parameters, as explained in the following.

Two main families of ANT architectures have been proposed in literature: *prediction-based* ANT [22, 23] and *reduced precision redundancy* (RPR) ANT [45, 46]. Hybrid approaches have also been studied in [46]. In the next sections, we briefly describe the features of the existing types of ANT, which mainly differ in the implementation of the estimator. Then, we concentrate on RPR ANT and provide a complete example of its application.

### 4.5.2 Prediction-Based ANT

In prediction-based ANT the estimator is a linear forward predictor, that provides an approximation ($Y_p$) of the correct output based on the recent history of the system. In particular, the estimate is based on a linear combination of the last $N_p$ outputs produced by the MDSP:

$$y_p[n] = \sum_{k=1}^{N_p} h_p[k]y[n-k]. \tag{4.25}$$

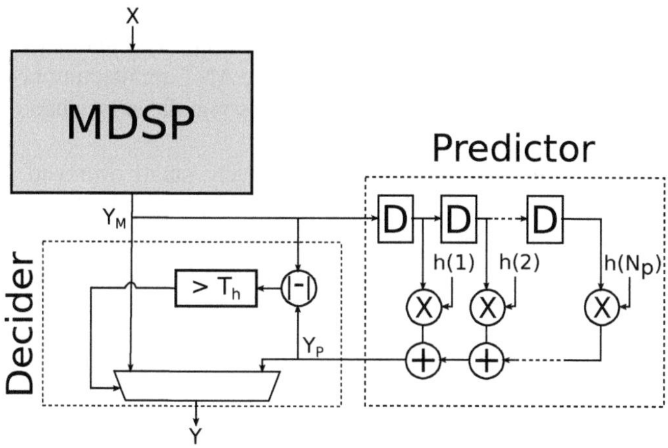


**Fig. 4.8** Prediction-based ANT architecture

The weights $h_p[k]$ are chosen to minimize the MSE of the predictor $\mathbb{E}[e_p^2[n]]$ in absence of VOS-induced timing violations (i.e., when $e_p[n] = \|y_{M,o}[n] - y_p[n]\|$). The way in which these coefficients are obtained is discussed in detail in [43].

The hardware that implements an estimator with the transfer function of Eq. (4.25) is shown in Fig. 4.8, where the boxes labeled with a "D" are delay elements, i.e. flip-flops. The figure also reports one of the possible decision schemes for this architecture, which follows the same principle of Eq. (4.19). The decider computes internally the prediction error and compares it to a threshold $T_h$. Given that prediction coefficients are computed to minimize the approximation error in absence of VOS, if $e_p[n]$ is large it is assumed that a timing violation has occurred. In that case, the predictor output is used as approximation of the correct value. If instead the error is less than $T_h$, the MDSP output is selected. Other more complex decision schemes have been proposed in [23]; they are not reported here for brevity.

The overheads in terms of cost and power of the estimator in prediction-based ANT depend on the length of the prediction window $N_p$. The simplest version uses a single past value of the MDSP output; in this case, it is referred to as *difference-based* estimator, since the value that is compared with $T_h$ is simply the (possibly weighted) difference between the current and previous values of $Y_M$. Clearly, a longer prediction window allows to obtain a more accurate estimate.

Setting the threshold $T_h$ for the decider is one of the main design issues of this type of architecture. In [23] the authors propose to use a multiple of the standard deviation of the prediction error ($T_h = k\sigma_{e_p}$). However, notice that this does not guarantee that the decider always selects the MDSP output when the system is free of VOS-induced errors. If the MDSP output varies suddenly, causing a large error in the predictor, the decider may assume a delay error has occurred when in reality it has not. In general, prediction-based ANT systems have good detection capabilities when the outputs of the MDSP exhibit a strong *temporal correlation*, that is when subsequent values of $Y_M$ are correlated with each other.

Another limitation of this scheme is that the predictor output is biased by errors occurred in the previous $N_p$ instants. In this time window, if another error occurs, detection and output estimation capabilities of the ANT architecture become much less accurate. In particular, performance degrades significantly when *error bursts* happen at the MDSP output.

In summary, despite its simplicity and relatively small overhead, prediction-based ANT is limited to few MDSP, such as narrowband low-pass filters, that exhibit strong output correlation. Moreover, it does not allow to aggressively scale the supply voltage $V_{VOS}$, since its effectiveness is limited to situations in which delay errors do not accumulate in the prediction window.

### 4.5.3 Reduced-Precision Redundancy ANT

RPR ANT uses an alternative scheme for the error control block; the entire architecture is depicted in Fig. 4.9. In this case, the estimator is a *low-complexity replica* of the MDSP, obtained reducing the precision of internal operations. The replica receives the same inputs as the MDSP at every clock cycle, and produces a corresponding approximation of the main block results. Thanks to the smaller and thus faster hardware operands involved, the replica can be designed to be timing compliant in VOS conditions. Its internal logic is designed so that approximate results differ from accurate ones in the LSBs; therefore, the approximation error has a small magnitude. This allows the decider to select between MDSP and replica outputs based on the absolute value of their difference, as explained for the prediction-based architecture.

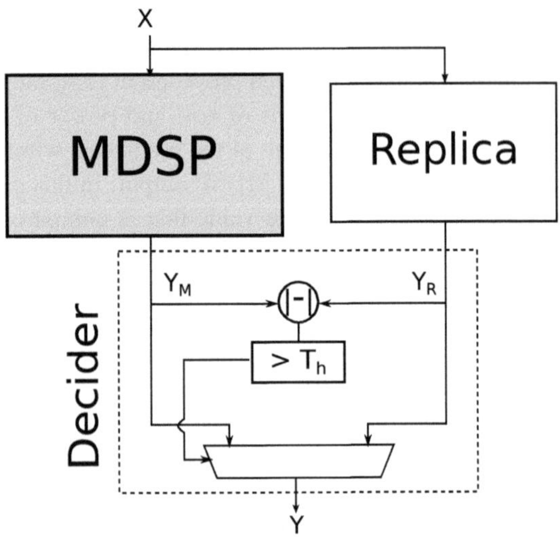


**Fig. 4.9** Reduced-precision redundancy ANT architecture

If the entire output images of MDSP and replica, i.e. the sets of output values produced by any sequence of inputs are known, the decision threshold $T_h$ can be computed as:

$$T_h = \max_{\forall \text{input}} \|y_{M,0}[n] - y_R[n]\| \tag{4.26}$$

where $y_{M,0}[n]$ and $y_R[n]$ are the error-free MDSP outputs and the replica outputs at time $n$, respectively. Equation (4.26) ensures that in absence of VOS-induced errors, the MDSP output is always selected by the decider. However, the entire images of $Y_M$ and $Y_R$ are easy to compute only for few simple MDSPs. Therefore, statistical thresholds similar to those proposed for prediction-based ANT might be necessary also in this case.

In RPR ANT the tradeoff between quality and EC block overheads is explored by varying the number of bits ($B_r$) used for the internal operands of the replica [which corresponds to the generic parameter **k** in Eq. (4.23)]. A larger replica consumes more power, both switching and leakage, and has a larger silicon cost. However, it produces a more accurate approximation of the MDSP output, and therefore it allows to maintain a higher quality at the global output $Y$. As explained in Sect. 4.5.1.2, the goal is typically to make errors due to VOS negligible with respect to external noise.

The advantage of RPR with respect to prediction-based schemes is its higher flexibility. In fact, since the output estimate is computed based only on the present value of inputs, this architecture behaves in an equivalent manner regardless of the correlation between subsequent outputs in the MDSP.[4] Also, the error mitigation capabilities of RPR ANT are not affected by recent errors, except for cases in which MDSP and replica have internal feedback networks. The drawback of RPR is that, especially for complex MDSPs, replicas have larger overheads with respect to linear predictors.

### 4.5.4 *Hybrid ANT*

Prediction-based ANT performance degrades when two timing errors occur too close in time, effectively limiting the minimum $V_{\text{vos}}$ at which the system can be operated. On the other hand, RPR ANT has a larger power overhead. A hybrid scheme, proposed in Fig. 4.10 can be used to overcome these limitations, exploiting the best features of both architectures.

[4]Different sequences of data may cause a different VOS-error rate, which results in a different quality, but they have no influence on the error mitigation capabilities of the architecture.

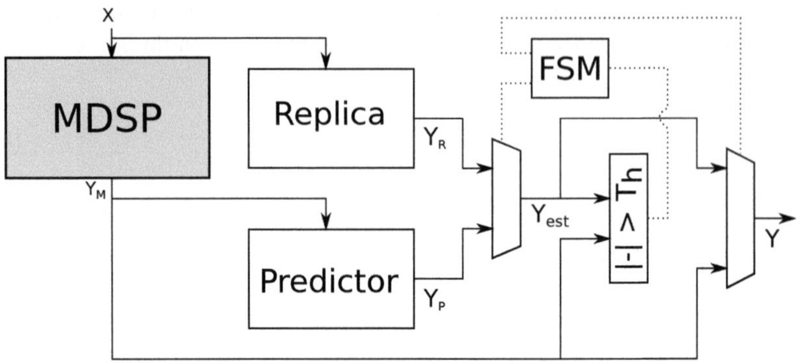


**Fig. 4.10** Hybrid ANT architecture

In hybrid ANT, the EC block is equipped with two different estimators: a linear predictor and a reduced-precision replica, designed as in Sects. 4.5.2 and 4.5.3, respectively. An additional finite state machine (FSM) selects between two operating modes. In normal conditions, the approximation of the correct output is obtained with the linear predictor ($Y_{\text{est}} = Y_P$) whereas the replica is turned off via power/clock gating, eliminating most of its overheads. When an error is sensed by the decider ($\|y_{\text{est}}[n] - y_M[n]\| > T_h$) the FSM turns on the replica and switches the leftmost multiplexer of Fig. 4.10, so that $Y_{\text{est}}$ becomes equal to $Y_R$. This preserves the error detection and mitigation capabilities of the EC block, which is not affected by the accumulation of errors in the predictor. The system remains in "RPR-mode" until no errors are detected for a given number of clock cycles. Then, it returns to normal operation. In [46] the authors show that hybrid ANT outperforms both RPR and prediction-based architectures in terms of output quality, for the particular case of a digital filter MDSP. The major drawback of this scheme is clearly the larger silicon area required.

## 4.6 A Case Study: RPR ANT Applied to a Finite Impulse Response Filter

To conclude the discussion on approximate computing for digital processing in smart systems, this section provides a detailed example of application of ANT, which as explained in Sect. 4.5, is one of the most promising approaches in this context. With a similar rationale, the hardware architecture that we optimize is a digital filter, due to its wide range of possible uses in smart systems applications (e.g., communication, multimedia, control, etc.) [43]. We select a 16th order low-pass FIR filter, with a cutoff frequency of $w_c = 0.1\pi$ rad/s.

The starting point for our optimization is the Verilog code that models the FIR filter at the register-transfer level (RTL). The particular implementation considered

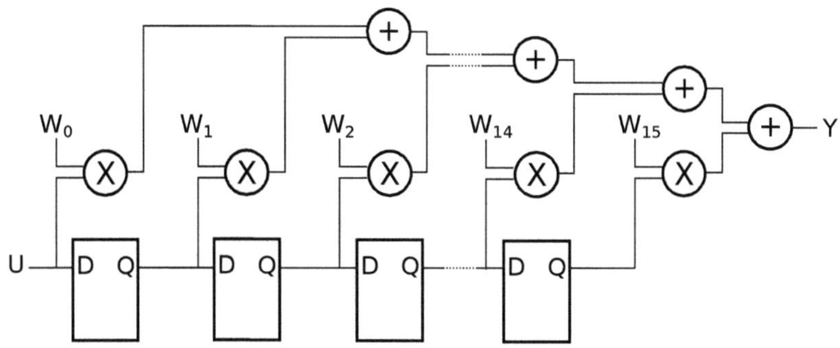


**Fig. 4.11** Direct form FIR filter block diagram

in this section uses the so-called direct form architecture, and its input ($U$) and output ($Y$) ports are 12-bit and 24-bit wide, respectively. A simplified block diagram of the FIR filter is shown in Fig. 4.11. Its transfer function is:

$$y[n] = \sum_{i=0}^{15} u[n-i] \cdot w[i]. \tag{4.27}$$

We adopt reduced-precision redundancy ANT as an optimization technique, because of its higher flexibility with respect to prediction-based approaches. In particular, we have anticipated that prediction-based ANT is effective only if the filter has a narrowband transfer function. RPR, on the contrary, can be exploited independently from the parameters of the FIR (tap-length and coefficient values), although the resulting performance may vary.

The objective in RPR ANT is to find the minimum VOS supply voltage and number of bits in the replica operands that satisfy a given quality constraint [as in Eq. (4.23)]. Previous approaches [45, 46] have provided a solid theoretical foundation for this type of optimization. However, their search for the optimal parameters is based on simplified and often unrealistic assumptions.

The most important limitation is related to the method of evaluation of VOS-induced errors and of their effects on the output quality. Specifically, the variance of errors introduced by ANT, expressed by Eq. (4.22), is a fundamental element for the computation of the SNR and many other quality metrics (e.g., MSE). This quantity depends both on the accuracy of approximations provided by the EC block, and on the rate at which the replica output is selected, over the entire time window considered.

In [45] the rate of VOS-induced errors is evaluated as the percentage of static timing paths in the MDSP that have a negative slack at a given $V_{vos}$. Since path activation depends on the sequence of input vectors, this method implicitly assumes that *all pairs of inputs* are equally probable. However, in most applications, the probability of occurrence of input vectors is not uniformly distributed (e.g., in arithmetic cores often the operands have a Gaussian distribution with zero mean) let alone that of input pairs. The assumption of uniformly probable path activation biases also the estimation of error significance, since every possible VOS-induced error is considered equally likely. This, in turn, affects the evaluation of which errors are detected and mitigated by the EC block, and of the corresponding approximation accuracy. All these aspects influence significantly Eq. (4.22), and might produce misleading results (both overly optimistic or pessimistic) on the number of replica operand bits required to ensure a minimum quality at a given $V_{\text{vos}}$.

Another important aspect that is partially neglected by literature is the effect of voltage scaling on the replica. We have mentioned that the replica must be timing compliant in VOS, in order to let the ANT system function correctly. However, this is not trivially achieved, and depends strongly on the topology of the circuit. It does not suffice to assume that the reduction in critical delay of the replica is proportional to the number of removed bits in its operands. Timing compliance must be checked accurately, and can also be enforced, for instance, with gate re-sizing. The latter technique, however, may increase the power overhead and reduce the total savings of the ANT system. All these details cannot be taken into account by optimization methods such as the one proposed in [45], which uses abstract high-level models of MDSP and replica.

To cope with these issues, we have developed a tool for the automatic generation of RPR ANT architectures on top of existing MDSPs. Our tool is based on *timing simulations* with input stimuli taken from the real application domain of the MDSP, which are used to evaluate the rate and impact of VOS-induced errors accurately. These simulations leverage a set of *standard cell library characterizations*, that model the target technology (usually CMOS) at different voltage points. This approach allows to consider non-uniform input/input-pairs distributions as well as secondary technological effects that have an impact on timing violations.

Moreover, the *replica implementation is obtained automatically from the MDSP netlist*, removing some inputs and propagating the simplifications in the internal logic. To make it timing compliant, the replica is then re-synthesized with state of the art commercial tools, changing gate sizing if needed. After synthesis, the power overheads are checked, and if they produce a negative power saving for the entire ANT system, the configuration is discarded. This procedure is repeated in a clever way, minimizing the number of synthesis and simulations required, until the optimal configuration in terms of VOS supply voltage and replica bits is found.

Using this tool, we have studied the application of RPR ANT to the FIR filter from an engineering perspective, gathering detailed results that were not obtainable with the methods originally proposed by literature. In the following sections, we first describe the experimental conditions, then we present some results that confirm the improvements achieved by our tool in terms of accuracy. In particular, we focus on the effects of input distributions and timing correlation on VOS-induced errors. To conclude, we show how the tool can be used to explore the power versus quality tradeoff.

### 4.6.1 Experimental Conditions

We performed experiments on the FIR filter targeting a 45 nm CMOS standard cell technology. The tool has been instructed to find the optimal VOS voltage in the range between 1.1 and 0.55 V, with a step of 0.05 V. The clock frequency was set to 250 MHz. This value allows to meet timing constraints exactly (with positive slack ~100 ps) for the single MDSP at $V_{dd} = 1.1$ V, which we considered as the *nominal* voltage condition. For the synthesis of MDSP and EC block we used Synopsys Design Compiler F-2011.09, while for timing and power analysis we used Synopsys PrimeTime Suite F-2011.12, and for behavioral and timing simulations Mentor Modelsim SE 6.4a. As quality metric, we used the SNR, defined as in Eq. (4.21) for all experiments. We measured the total power savings and area overheads of the RPR ANT architecture with the following equations:

$$P_{\text{sav}} = \left(1 - \frac{P_{\text{rpr}}(V_{\text{VOS}}, B_r)}{P_{\text{mdsp}}(V_{\text{nom}})}\right) \cdot 100\,\% \tag{4.28}$$

$$A_{\text{ovr}} = \left(\frac{A_{\text{rpr}}(V_{\text{VOS}}, B_r)}{A_{\text{mdsp}}} - 1\right) \cdot 100\,\% \tag{4.29}$$

where $V_{\text{nom}}$, $V_{\text{VOS}}$, $P_{\text{mdsp}}$, and $P_{\text{rpr}}$ are defined as in Sect. 4.5, $A_{\text{mdsp}}$ and $A_{\text{rpr}}$ represent the silicon area of the MDSP and replica, respectively, and $B_r$ is the number of bits in the replica inputs, which in turn determines the widths of internal operands.

### 4.6.2 Impact of Input Distribution and Correlation

The first set of experiments aims at highlighting the strong dependence of the distribution and temporal sequence of inputs on the rate of VOS-induced errors, and consequently on the performance of RPR ANT. For this purpose, we ran our optimization tool on the FIR filter four times, with identical settings, but using four different sets of input vectors (S1–S4) in timing simulations. The first two sets contain uniformly distributed vectors. In S1, vectors are in a random sequence, while in S2 they are *sorted* numerically, in order to enforce a strong temporal correlation in the MSBs (with a large probability, only the LSBs vary between two consecutive vectors). S3 and S4 instead represent more realistic inputs for a digital filter. They contain two full swing sinusoids, respectively, at $w_{s3} = 0.01\pi$ rad/s, that is one tenth of the cutoff frequency, and $w_{s4} = w_c = 0.1\pi$ rad/s. In all cases, we suppose that input data is noiseless; we measure quality in terms of SNR, but considering only VOS-induced errors, and we arbitrarily set the minimum quality requirement to 25 dB. Moreover, we limit our analysis to those VOS voltages that produce an error rate in the MDSP (computed as number of timing violations over total number of vectors) greater than 0 % and smaller than 20 %.

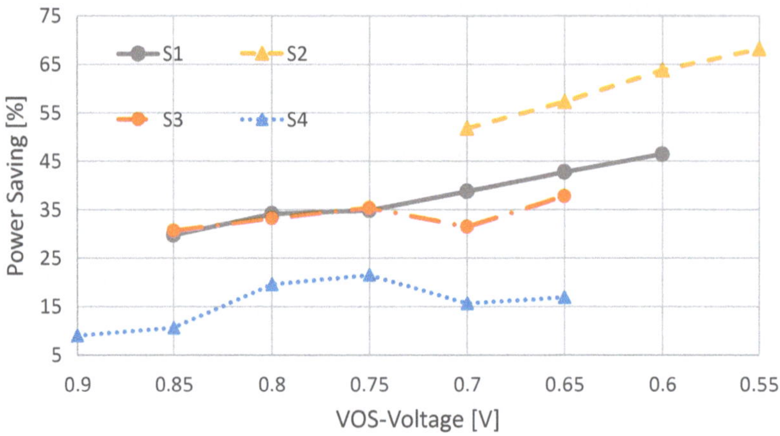


**Fig. 4.12** RPR ANT power saving vs. $V_{vos}$ for different input sets

Figure 4.12 reports the maximum power savings obtained at each $V_{vos}$ point with the mentioned constraints. The graph clearly shows the impact of different input vectors on the effectiveness of RPR ANT. Two phenomena are particularly important. First of all, at a given $V_{vos}$ the error rates are significantly different depending on the input set. As a numerical example, at $V_{vos} = 0.7$ V, the error rates in S2 and S4 are 0.05 % and 12 %, respectively. Consequently, depending on the input set, the replica output is selected more or less frequently, and hence a different value of $B_r$ is required to achieve the desired quality. At $V_{vos} = 0.7$ V, the errors for S2 are so rare that a replica that only considers the 2 MSBs of the filter input for its computations is sufficient; in the case of S4, instead, 6 bits are required to achieve 25 dB of SNR. At the end, the noticeable product of these phenomena is that the power savings in the two cases differ by almost 40 %. Obviously, the input sets used in this section constitute an extreme example. However, they serve as a mean to demonstrate that simplified assumptions on VOS-induced errors as those in [46] may lead to unrealistic results.

Notice that Fig. 4.12 also indirectly shows one of the features that make ANT effective. We mentioned that the positive slack of the MDSP in nominal conditions is only 100 ps, meaning that the clock frequency is appropriate (not too small), according to a classic worst-case design paradigm. However, we discovered that for voltages between 1.05 and 0.9 V, none of the input sets produces a timing violation in the MDSP, i.e. the error rate remains 0 %. This means that the critical timing path of the FIR, which corresponds to the cascade of carry chains in all adders and multipliers (see Fig. 4.11), is never excited by the provided stimuli. This is an evident reason why relaxing the 100 % accuracy constraint, working in VOS, and limiting the impact of the rare errors with ANT is a very effective way to reduce power.

### 4.6.3 Power Versus Quality Tradeoff

In this section, we demonstrate a more realistic use of our tool for the optimization of the FIR filter, showing how it can be leveraged to explore the quality versus power tradeoff. To do so, we select a popular application of hardware FIRs, which has been also considered in [46], that is the low-pass filtering at the receiver of a QPSK communication system. In particular, we instruct our optimization tool to perform simulations with a set of inputs generated from a MATLAB model of the IEEE 802.11g WiFi standard, which includes QPSK among the possible modulations. We assume that the inputs to the FIR are affected by additive white Gaussian noise (AWGN) due to channel interferences. As explained in [46], a minimum SNR at the filter output of 21.5 dB ensures a bit error rate of $10^{-7}$. Therefore, we add AWGN to the input set, so that the output SNR of the filter is approximately at such value. Then, we let the tool generate all feasible replica configurations at each $V_{\text{VOS}}$, without imposing a quality constraint.[5] In this context, feasible means that the ANT architecture has positive power savings, and that the replica is timing compliant. The results on a power saving versus quality plane are reported in Fig. 4.13, in which each curve corresponds to a value of the VOS supply voltage, and points on it correspond to different replica input widths ($B_r$).

Two trends can be identified in the graph. On the left side, quality is determined by the channel AWGN noise, and saturates to a horizontal line. Therefore, all versions of the ANT architecture have the same final quality, independently from $B_r$. On the right side of the graph, instead, the errors due to VOS become dominant with respect to channel noise, and the curves assume a typical *Pareto* shape: implementations with a smaller $B_r$ have smaller overheads, and hence higher total power savings, but also a worse quality. A quality threshold can be visualized as a horizontal line on the graph. The optimal RPR architecture for that quality is found as the rightmost point among all those above the line.

If a designer wants to maintain a quality of 21.5 dB at the output of the FIR filter, in order to respect the constraints imposed by the 802.11g standard, RPR allows to save up to 44.96 % of *total* power. This is achieved by setting the voltage to $V_{\text{VOS}} = 0.60\,\text{V}$, and inserting in the EC block a replica that considers the 6 MSBs of the 12-bit input for its internal computations. The total area overhead of the EC block, including replica and decider is 88.39 %. Notice that this significant power saving is obtained *without any effective impact on quality*. In fact, quality was already bounded by the noise on the channel.

Interestingly, Fig. 4.13 shows that a lower $V_{\text{VOS}}$ does not always correspond to a larger power saving, for a given quality constraint. This happens because of the gate re-sizing performed by our tool in order to make the replica timing compliant. In fact, at lower $V_{\text{VOS}}$, faster gates, which are also larger and more consuming, are needed to meet timing, and the resulting design often consumes more despite the

[5] In its normal operating mode, the tool only synthesizes the replica configuration with minimum $B_r$ to satisfy a quality constraint at each $V_{\text{VOS}}$.

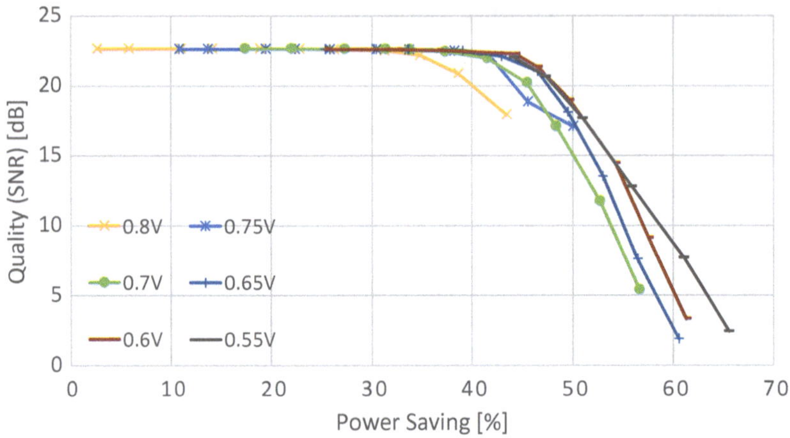


**Fig. 4.13** RPR ANT quality vs. power saving tradeoff for a 16th order FIR filter

reduced supply voltage. This trend shows that accurate timing and power evaluations for the replica, which were not performed in [46], are mandatory to correctly estimate the effectiveness of RPR. Notice that the same phenomenon is visible also in Fig. 4.12, since power saving does not increase monotonically with the reduction of $V_{vos}$ for some input sets.

## 4.7 Conclusions

In this chapter, we have surveyed the most popular design solutions based on the approximate computing paradigm. We have shown how, despite its relatively recent formalization, AC has stimulated interesting research at all abstraction levels, from single transistors to complete systems and to software. The growing interest on this subject is demonstrated by the fact that most cited works have been published in the last 2 or 3 years. Therefore, it is easy to foresee that in the near future, even more effort from will be devoted to the progress of this field of research by academia and industry.

We envision approximate computing as an effective way to reduce power and energy consumption in smart systems, providing as motivation the fact that applications performed by these devices are often error resilient. As an example, we have applied a popular AC technique to the optimization of a digital filter, which is one of the most commonly found hardware modules in digital signal processing applications. We have considered the situation in which the filter is part of the receiver of a digital communication system, affected by noise on the channel, and we have shown how, in this setting, almost 50 % power saving can be achieved without an effective impact on output quality. Moreover, higher power savings can be obtained if a partial quality reduction is accepted.

## References


1. R. Abdallah, N. Shanbhag, Minimum-energy operation via error resiliency. IEEE Embed. Syst. Lett. **2**(4), 115–118 (2010)
2. A. Alaghi, J.P. Hayes, Survey of stochastic computing. ACM Trans. Embed. Comput. Syst. **12**(2s), 92:1–92:19 (2013)
3. N. Banerjee, G. Karakonstantis, K. Roy, Process variation tolerant low power DCT architecture, in *Design, Automation Test in Europe Conference Exhibition, DATE '07* (2007), pp. 1–6
4. N. Bombieri, D. Drogoudis, G. Gangemi, R. Gillon, E. Macii, M. Poncino, S. Rinaudo, F. Stefanni, D. Trachanis, M. van Helvoort, SMAC: smart systems co-design, in *Euromicro Conference on Digital System Design (DSD)* (2013), pp. 253–259
5. S. Chakradhar, A. Raghunathan, Best-effort computing: re-thinking parallel software and hardware, in *47th ACM/IEEE Design Automation Conference (DAC)* (2010), pp. 865–870
6. L.N.B. Chakrapani, K.V. Palem, A probabilistic boolean logic for energy efficient circuit and system design, in *Proceedings of the 2010 Asia and South Pacific Design Automation Conference, ASPDAC '10* (IEEE Press, Piscataway, NJ, 2010), pp. 628–635
7. J. Chen, J. Hu, Energy-efficient digital signal processing via voltage-overscaling-based residue number system. IEEE Trans. Very Large Scale Integr. Syst. **21**(7), 1322–1332 (2013)
8. V. Chippa, D. Mohapatra, A. Raghunathan, K. Roy, S. Chakradhar, Scalable effort hardware design: exploiting algorithmic resilience for energy efficiency, in *47th ACM/IEEE Design Automation Conference (DAC)* (2010), pp. 555–560
9. V. Chippa, A. Raghunathan, K. Roy, S. Chakradhar, Dynamic effort scaling: managing the quality-efficiency tradeoff, in *48th ACM/EDAC/IEEE Design Automation Conference (DAC)* (2011), pp. 603–608
10. V. Chippa, S. Chakradhar, K. Roy, A. Raghunathan, Analysis and characterization of inherent application resilience for approximate computing, in *50th ACM/EDAC/IEEE Design Automation Conference (DAC)* (2013), pp. 1–9
11. J. Cong, V. Sarkar, G. Reinman, A. Bui, Customizable domain-specific computing. IEEE Trans. Des. Test Comput. **28**(2), 6–15 (2011)
12. M. Crepaldi, M. Grosso, A. Sassone, S. Gallinaro, S. Rinaudo, M. Poncino, E. Macii, D. Demarchi, A top-down constraint-driven methodology for smart system design. IEEE Circuits Syst. Mag. **14**(1), 37–57 (2014)
13. S. Das, C. Tokunaga, S. Pant, W.H. Ma, S. Kalaiselvan, K. Lai, D. Bull, D. Blaauw, Razorii: in situ error detection and correction for PVT and SER tolerance. IEEE J. Solid State Circuits **44**(1), 32–48 (2009)
14. 2014 MultiAnnual Strategic Research and Innovation Agenda (MASRIA) for the ECSEL Joint Undertaking (2014)
15. D. Ernst, S. Das, S. Lee, D. Blaauw, T. Austin, T. Mudge, N.S. Kim, K. Flautner, Razor: circuit-level correction of timing errors for low-power operation. IEEE Micro **24**(6), 10–20 (2004)
16. B.R. Gaines, Stochastic computing, in *Proceedings of the Spring Joint Computer Conference, AFIPS '67* (1967), pp. 149–156
17. S. Ghosh, S. Bhunia, K. Roy, Crista: a new paradigm for low-power, variation-tolerant, and adaptive circuit synthesis using critical path isolation. IEEE Trans. Comput. Aided Des. Integr. Circuits Syst. **26**(11), 1947–1956 (2007)
18. W. Gross, V. Gaudet, A. Milner, Stochastic implementation of LDPC decoders, in *Conference Record of the Thirty-Ninth Asilomar Conference on Signals, Systems and Computers* (2005), pp. 713–717
19. V. Gupta, D. Mohapatra, A. Raghunathan, K. Roy, Low-power digital signal processing using approximate adders. IEEE Trans. Comput. Aided Des. Integr. Circuits Syst. **32**(1), 124–137 (2013)
20. J. Han, M. Orshansky, Approximate computing: an emerging paradigm for energy-efficient design, in *18th IEEE European Test Symposium (ETS)* (2013), pp. 1–6

21. K. He, A. Gerstlauer, M. Orshansky, Controlled timing-error acceptance for low energy idct design, in *Design, Automation Test in Europe Conference Exhibition (DATE)* (2011), pp. 1–6
22. R. Hegde, N. Shanbhag, Energy-efficient signal processing via algorithmic noise-tolerance, in *International Symposium on Low Power Electronics and Design (ISLPED)* (1999), pp. 30–35
23. R. Hegde, N. Shanbhag, Soft digital signal processing. IEEE Trans. Comput. Aided Des. Integr. Circuits Syst. **9**(6), 813–823 (2001)
24. J. Huang, J. Lach, G. Robins, A methodology for energy-quality tradeoff using imprecise hardware, in *49th ACM/EDAC/IEEE Design Automation Conference (DAC)* (2012), pp. 504–509
25. A. Kahng, S. Kang, Accuracy-configurable adder for approximate arithmetic designs, in *49th ACM/EDAC/IEEE Design Automation Conference (DAC)* (2012), pp. 820–825
26. A. Kahng, S. Kang, R. Kumar, J. Sartori, Slack redistribution for graceful degradation under voltage overscaling, in *15th Asia and South Pacific Design Automation Conference (ASP-DAC)* (2010), pp. 825–831
27. G. Karakonstantis, K. Roy, Voltage over-scaling: a cross-layer design perspective for energy efficient systems, in *20th European Conference on Circuit Theory and Design (ECCTD)* (2011), pp. 548–551
28. G. Karakonstantis, D. Mohapatra, K. Roy, System level DSP synthesis using voltage overscaling, unequal error protection and adaptive quality tuning, in *IEEE Workshop on Signal Processing Systems (SiPS)* (2009), pp. 133–138
29. P. Kulkarni, P. Gupta, M. Ercegovac, Trading accuracy for power with an underdesigned multiplier architecture, in *24th International Conference on VLSI Design (VLSI Design)* (2011), pp. 346–351
30. K.Y. Kyaw, W.L. Goh, K.S. Yeo, Low-power high-speed multiplier for error-tolerant application, in *IEEE International Conference of Electron Devices and Solid-State Circuits (EDSSC)* (2010), pp. 1–4
31. L. Leem, H. Cho, J. Bau, Q. Jacobson, S. Mitra, ERSA: error resilient system architecture for probabilistic applications, in *Design, Automation Test in Europe Conference Exhibition (DATE)* (2010), pp. 1560–1565
32. J. Liang, J. Han, F. Lombardi, New metrics for the reliability of approximate and probabilistic adders. IEEE Trans. Comput. **62**(9), 1760–1771 (2013)
33. C. Liu, J. Han, F. Lombardi, A low-power, high-performance approximate multiplier with configurable partial error recovery, in *Proceedings of the Conference on Design, Automation & Test in Europe (DATE)* (2014), pp. 95:1–95:4
34. S.L. Lu, Speeding up processing with approximation circuits. Computer **37**(3), 67–73 (2004)
35. E. Macii, *Ultra Low-Power Electronics and Design* (Springer US, New Mexico, NM, 2004)
36. H. Mahdiani, A. Ahmadi, S. Fakhraie, C. Lucas, Bio-inspired imprecise computational blocks for efficient VLSI implementation of soft-computing applications. IEEE Trans. Circuits Syst. Regul. Pap. **57**(4), 850–862 (2010)
37. J. Meng, S. Chakradhar, A. Raghunathan, Best-effort parallel execution framework for recognition and mining applications, in *IEEE International Symposium on Parallel Distributed Processing* (2009), pp. 1–12
38. D. Mohapatra, G. Karakonstantis, K. Roy, Significance driven computation: a voltage-scalable, variation-aware, quality-tuning motion estimator, in *Proceedings of the ACM/IEEE International Symposium on Low Power Electronics and Design (ISLPED)* (2009), pp. 195–200
39. D. Mohapatra, V. Chippa, A. Raghunathan, K. Roy, Design of voltage-scalable meta-functions for approximate computing, in *Design, Automation Test in Europe Conference Exhibition (DATE)* (2011), pp. 1–6
40. K. Nepal, Y. Li, R.I. Bahar, S. Reda, Abacus: a technique for automated behavioral synthesis of approximate computing circuits, in *Proceedings of the Conference on Design, Automation & Test in Europe (DATE)* (2014), pp. 361:1–361:6
41. K. Palem, A. Lingamneni, What to do about the end of Moore's law probably, in *Proceedings of the 49th Design Automation Conference (DAC)* (2012), pp. 924–929

42. W.J. Poppelbaum, C. Afuso, J.W. Esch, Stochastic computing elements and systems, in *Proceedings of the Joint Computer Conference (AFIPS)* (1967), pp. 635–644
43. J.G. Proakis, D.G. Manolakis, *Digital Signal Processing*. Principles, Algorithms, and Applications, 3rd edn. (Prentice-Hall, Princeton, NJ, 1996)
44. A. Ranjan, A. Raha, S. Venkataramani, K. Roy, A. Raghunathan, ASLAN: synthesis of approximate sequential circuits, in *Design, Automation and Test in Europe Conference and Exhibition (DATE)* (2014), pp. 1–6
45. B. Shim, N. Shanbhag, Performance analysis of algorithmic noise-tolerance techniques, in *Proceedings of the 2003 International Symposium on Circuits and Systems (ISCAS)*, vol. 4 (2003), pp. IV-113–IV-116
46. B. Shim, S. Sridhara, N. Shanbhag, Reliable low-power digital signal processing via reduced precision redundancy. IEEE Trans. Very Large Scale Integr. Syst. **12**(5), 497–510 (2004)
47. D. Shin, S. Gupta, Approximate logic synthesis for error tolerant applications, in *Design, Automation Test in Europe Conference Exhibition (DATE)* (2010), pp. 957–960
48. D. Shin, S. Gupta, A new circuit simplification method for error tolerant applications, in *Design, Automation Test in Europe Conference Exhibition (DATE)* (2011), pp. 1–6
49. A. Tanenbaum, *Computer Networks*, 4th edn. Prentice Hall Professional Technical Reference (Prentice Hall, Princeton, NJ, 2002)
50. S. Venkataramani, A. Sabne, V. Kozhikkottu, K. Roy, A. Raghunathan, Salsa: systematic logic synthesis of approximate circuits, in *49th ACM/EDAC/IEEE Design Automation Conference (DAC)* (2012), pp. 796–801
51. S. Venkataramani, V.K. Chippa, S.T. Chakradhar, K. Roy, A. Raghunathan, Quality programmable vector processors for approximate computing, in *Proceedings of the 46th IEEE/ACM International Symposium on Microarchitecture (MICRO-46)* (2013), pp. 1–12
52. S. Venkataramani, S. Chakradhar, K. Roy, A. Raghunathan, Approximate computing for efficient information processing, in *12th IEEE Symposium on Embedded Systems for Real-time Multimedia (ESTIMedia)* (2014), pp. 9–10
53. A. Verma, P. Brisk, P. Ienne, Variable latency speculative addition: a new paradigm for arithmetic circuit design, in *Design, Automation and Test in Europe (DATE)* (2008), pp. 1250–1255
54. Z. Yang, A. Jain, J. Liang, J. Han, F. Lombardi, Approximate XOR/XNOR-based adders for inexact computing, in *13th IEEE Conference on Nanotechnology (IEEE-NANO)* (2013), pp. 690–693
55. Y. Yetim, M. Martonosi, S. Malik, Extracting useful computation from error-prone processors for streaming applications, in *Proceedings of the Conference on Design, Automation and Test in Europe (DATE)* (2013), pp. 202–207
56. N. Zhu, W.L. Goh, K.S. Yeo, An enhanced low-power high-speed adder for error-tolerant application, in *Proceedings of the 12th International Symposium on Integrated Circuits (ISIC)* (2009), pp. 69–72
57. N. Zhu, W.L. Goh, W. Zhang, K.S. Yeo, Z.H. Kong, Design of low-power high-speed truncation-error-tolerant adder and its application in digital signal processing. IEEE Trans. Very Large Scale Integr. Syst. **18**(8), 1225–1229 (2010).
58. N. Zhu, W.L. Goh, G. Wang, K.S. Yeo, Enhanced low-power high-speed adder for error-tolerant application, in *International SoC Design Conference (ISOCC)* (2010), pp. 323–327
59. N. Zhu, W.L. Goh, K.S. Yeo, Ultra low-power high-speed flexible probabilistic adder for error-tolerant applications, in *International SoC Design Conference (ISOCC)* (2011), pp. 393–396

# Chapter 5
# Discrete Power Devices and Power Modules

**Aleš Chvála, Davide Cristaldi, Daniel Donoval, Giuseppe Greco, Juraj Marek, Marián Molnár, Patrik Príbytný, Angelo Raciti, and Giovanni Vinci**

## 5.1 Introduction

The purpose of this chapter is to provide to the designers a first-level entry to a flexible software ambient for some tasks related to the design, integration, and co-simulation of smart subsystems/components. In particular, the objective is to create and build models that will make the design and optimization of smart systems more efficient by enabling multi-domain simulations at various design levels for the variety of devices found in this kind of systems. The models are built according to the specific requirements defined in a few of application fields. The aim is to simplify the modeling-related obstacles affecting multi-domain system-level simulations. Within each case-study, a model is developed according to the type of components and subsystems. In particular, the activities are related to the modeling of discrete power devices and power modules and to the interactions between multiple physical domains, like electrical and thermal ones, which are key issues looking to obtain more and more efficient and reliable devices.

In the first part is developed an innovative methodology having as a main objective the implementation of an integrated power electronics module (IPEM) physics-based macro-model. The main aim is the co-simulation of the thermal and

A. Chvála • D. Donoval • J. Marek • M. Molnár • P. Príbytný
Slovak University of Technology in Bratislava, Bratislava, Slovakia
e-mail: ales.chvala@stuba.sk; daniel.donoval@stuba.sk; juraj.marek@stuba.sk; marian.molnar@stuba.sk; patrik.pibytny@stuba.sk

D. Cristaldi • A. Raciti (✉)
University of Catania, Catania, Italy
e-mail: davide.cristaldi@dieei.unict.it; angelo.raciti@dieei.unict.it

G. Greco • G. Vinci
STMicroelectronics, Catania, Italy
e-mail: giuseppe.greco@st.com; giovanni.vinci@st.com



N. Bombieri et al. (eds.), *Smart Systems Integration and Simulation*,
DOI 10.1007/978-3-319-27392-1_5

electrical phenomena. In the second part, the analysis of both a low-voltage vertical-power MOSFET failure during UIS and a DC–DC converter is performed in order to better predict the physical parameters of the power MOSFET model and the converter.

## 5.2 Power Modules Analysis, Modeling, and Simulation

Unlike past decades, when severe criteria on reliability of power electronics were mainly prerogative of industrial and military context, nowadays, several more common fields of applications require power electronics equipment able to manage significant amount of power with low failure rates. The growth of sectors like hybrid automotive traction or domestic renewable energy plants, pushed electronics players to develop solutions able to satisfy all requirements requested by system hosts. More and more integrated electronics systems push the hosted modules to have high-power density, reduced volume, endurance to mechanical stress, and suitable management of the generated heat. IPEMs well represent a good example among all the mentioned requirements, and their production nowadays are moving huge amount of investments for equipment among the manufacturing companies.

As these IPEMs are confined into small volumes in relation to the power that they have to handle, it is necessary to address temperature management issues. Furthermore, as the integrated power modules have to operate in environments characterized by temperatures that can exceed 100 °C, the managing of heat flow becomes a crucial aspect that has to be evaluated already at the design level. The availability of accurate electrothermal simulation models, able to predict the evolutions of dynamics in real operating conditions, provides the advantage to correctly face electrical, thermal, and mechanical aspects at the design phase [1–3]. This aspect leads to a reduced number of prototypes necessary for managing the characterization experimental tests. The top view of an IPEM is shown in Fig. 5.1.

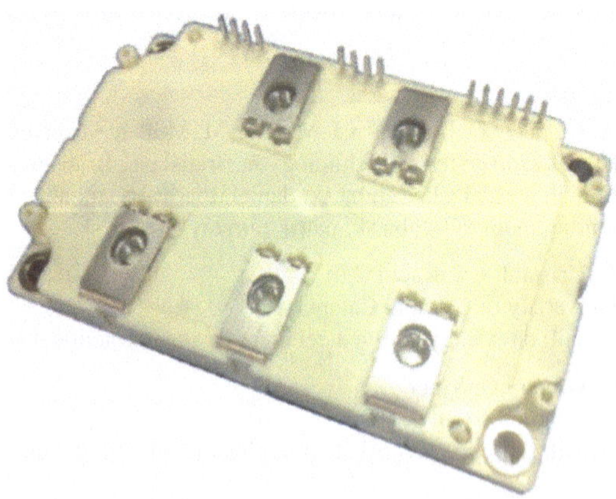

**Fig. 5.1** Top view of an IPEM

A few of activities related to IPEMs electrothermal modeling focus on steady-state and transient thermal analyses of multi-chip power devices, either adopting finite element approaches (FEAs) [4, 5] or boundary element based methods (BEMs) [6–8]. In recent years, research activities were mainly addressed to assemblies of packaged discrete power devices since IPEMs were not yet well-recognized as stand-alone modules. Novel modeling approaches, having the specific purpose to obtain an electrothermal model of IPEM are now appearing. In such cases some techniques try to analyze the IPEM physics with the purpose to model each domain in a separate way [9].

The understanding of the thermal behavior of the device is essential and the extraction of the thermal impedance matrix is a quite common strategy that is extracted either from experimental measurements [10, 11] or by refined three-dimensional thermal simulations. Among the latter type of simulations, the numerical computation approach through finite element methods (FEMs) is the most diffused [12]. More discussion on the state-of-the-art may be found by interested readers in [13], where the advantages and drawbacks of the available software packages and methodologies are recalled.

In the case of IPEMs the problem is the modeling of electrical and thermal systems featuring large different times to reach the steady-state behaviors. The methodology described in the present chapter allows overcoming many limitations encountered by using the state-of-the-art techniques. It is an alternative strategy leading to an electrothermal circuit model that may be directly used in a standard PSpice-like platform. Moreover, the part of the IPEM model, which is related to the thermal modeling, is generated through a customized EDA flow able to derive an equivalent PSpice-like netlist [14]. Finally, effects on the power losses due to non-ideal interconnections (parasitic phenomena) are taken into account by including in the implementation of the electric circuit the specific lumped parameters. Finally, the accuracy and effectiveness of the generated model have been verified by using, as a benchmark, accurate simulation results obtained by FEM packages.

### *5.2.1 Physical Domains Involved*

During the working operations, electrical, thermal, and mechanical aspects have to be accounted in order to ensure high reliability and to avoid the breakdown failure. In fact, the power losses generated into the devices are responsible for temperature swings, which produce mechanical stress, especially in soldering, due to the mismatch between the different coefficients of the thermal expansion of the materials in the IPEM. In order to investigate the issues related to the various domains, different software packages are necessary, with the resulting need of exchange of data, which often occurs manually, between them.

In order to overcome these limitations, a new strategy based on a standard PSpice-like electrothermal model specifically thought for IPEMs has been developed. The main advantage of the proposed approach relies on the possibility of

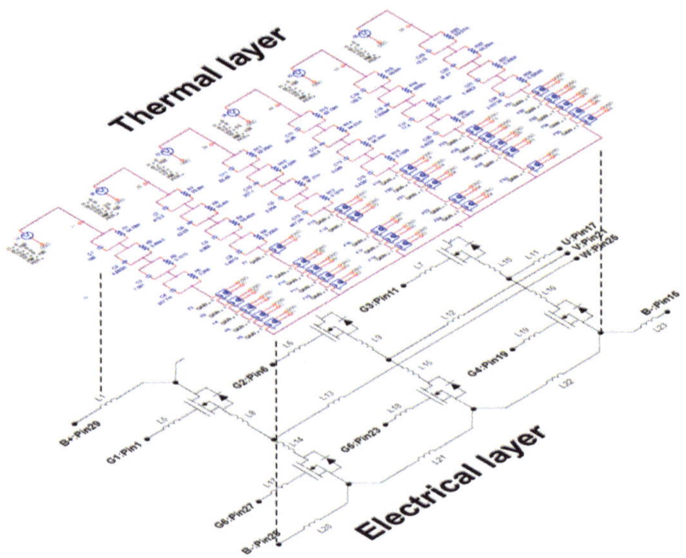


**Fig. 5.2** Layered approach scheme adopted for the IPEM electrothermal modeling

reducing a multi-domain electrothermal model to a pure single-domain electrical one. This result was achieved by adopting a mapping where the thermal quantities are treated as equivalent electrical ones. The generated model is conceived as a stack of two layers, respectively, representing the thermal and the electrical contributes (Fig. 5.2). The thermal layer represents the most complex part to be generated, and its implementation passes from the processing of a series of simulations results produced by FEM analyses. In fact, in order to properly evaluate the temperature rise due to the power losses, the knowledge of the self-heating and cross-heating effects is necessary.

Electrical and thermal layers are linked together by self-heating active layer, capable of accepting the junction temperatures as inputs in order to take into account the dependence on the temperature of some basic parameters of the devices like $V_{GE(th)}$, $V_{CE(sat)}$, or $BV_{CES}$.

### 5.2.2 *Links with the Software Platform and Abstraction Domain Matrix*

The design flow of the IPEM starts in an FEM environment (COMSOL Multiphysics). Once all aspects of the FEM analysis have been defined (geometry, thermal load, boundary conditions), the post-processing thermal impedance curves are exported. With the aid of a fitting tool software (curve fitting tool of MATLAB in this case), the curves are processed and the values of resistance and capacitance are estimated in order to build the electrical equivalent circuit topology of the IPEM thermal model [15, 16]. Hence, the thermal model is in the form of electrical circuit and it is simulated together with the electrical layer (active and parasitic layers) in a pure PSpice-like environment. The whole methodology proposed is described in following sections.

With the use of abstraction tools, like WiCkeD or TRAPPIST, starting from low-level models (PSpice), the corresponding Verilog-A model can be created in order to be used into the Keysight's advanced design system (ADS). The final simulation may be carried out in SystemVue. The part of the software platform covered by this chapter is shown in Fig. 5.3.

### *5.2.3 Model Discrepancies from the Compliance with Specific User Requirements*

The requirements (a–d) for discrete power devices and power modules, and the features of the IPEM macro-model (•) that are here presented, are discussed below.

(a) At system level there should be a mechanism to distribute the ambient temperature to all subparts of the design. To study the influence of the temperature on the system's performance it should be easy to set up multiple simulations at different temperature.

- The ambient temperature is not distributed to all the subparts of the system but it is used as reference value in order to calculate the junction temperature. Parametric studies can be easily performed by setting different values of the ambient temperature.

(b) The platform should be able to simulate behavioral models written in Verilog-A. If such a model defines the working temperature of other subparts of the system, the simulation platform should support the propagation of the temperature value to the other subparts.

- Not implemented. The model doesn't define the working temperatures of other subparts of the system beyond the ambient temperature one.

(c) Temperature variations typically occur at a much lower rate than electrical signals. Therefore, it should be possible to co-simulate the thermal model at a different time scale as the other parts of the system.

- By converting the instantaneous power losses in average power losses and by applying this latter to the thermal layer represented by a PSpice-like circuit, it is possible to have comparable simulation times for both the electrical and thermal layers simulations. However, it is required an iterative method that implies a segregation of the electrical and thermal simulations and exchange of data between the two layers. In more detail, the two analyses are

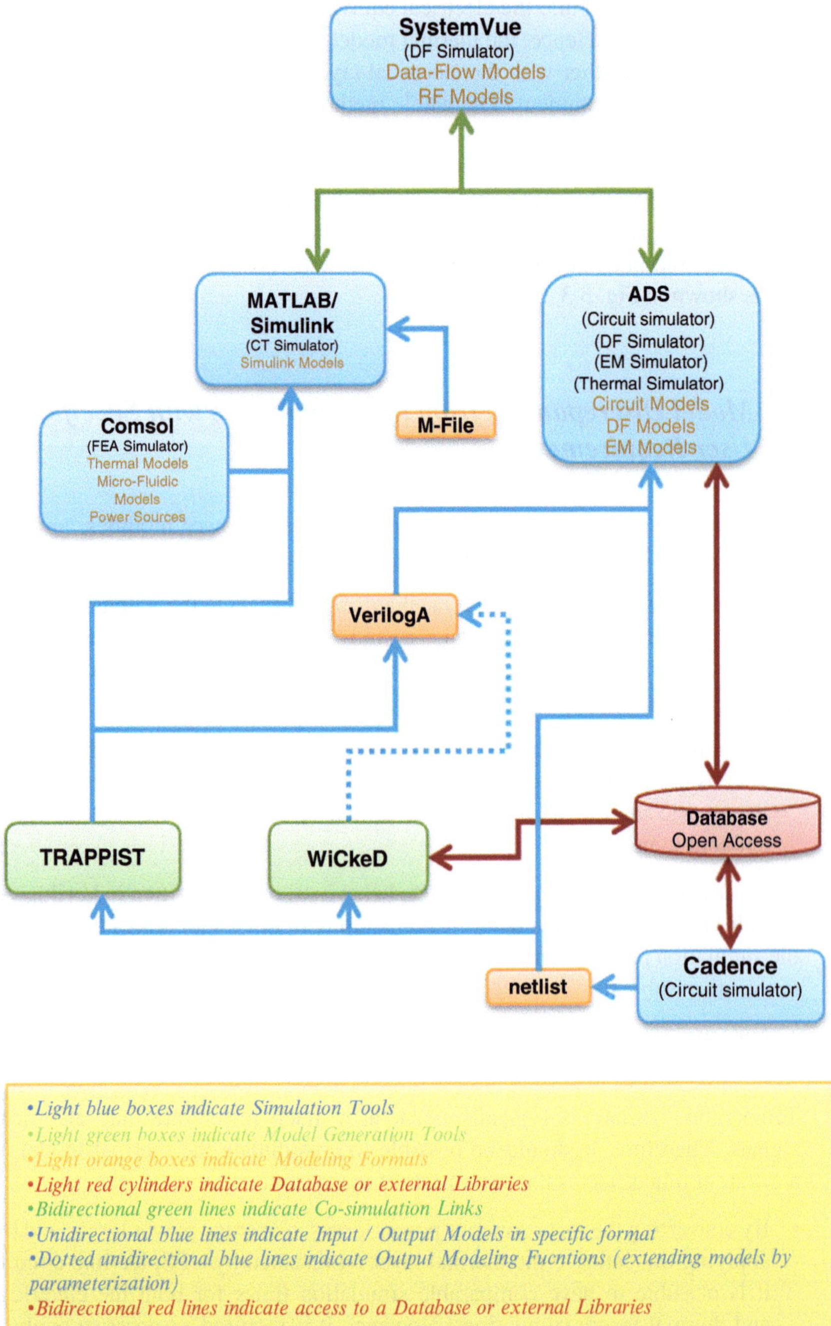


**Fig. 5.3** Design flow in the software platform

performed in separate ways and, iteratively, power losses and temperatures values are exchanged between the layers until the error on the temperature variations is below a prefixed threshold (convergence condition).

(d) Through the platform it should be possible to implement a full electrical–thermal model of a power system module by providing a final spice-like macro-model. This model will be logically composed by three main parts: an electrical spice-like layer containing the passive parasitic components, a thermal–electrical spice-like equivalent layer that will take into account the temperature propagation inside the system, and a third "linking" layer composed by self-heating active power device models that will connect the two previous layers through a power/temperature feedback.

- The electrothermal model is a full PSpice-like model composed by three layers: the parasitic layer and the thermal layer that are connected by the self-heating active layer. Through the platform it is possible to implement the whole system as a full electrical–thermal model, that in turns it is a macro-model represented by lumped-parameter circuits.

### *5.2.4 IPEM Electrothermal Analysis*

Smart systems are present in several fields of application and the IPEM represents a typical element that often is a key part of novel smart systems built in the field of power electronics applications. Typically, power management modules are controlled by sensors that provide, through feedback loops, controlling data to analog or digital drivers with the objective of optimizing the power flow to critical load by guarantying stability in terms of electrical and thermal response of the IPEM. Three different application domains may be identified for representing IPEM modules: discrete and power devices, analog and RF, MEMS sensors and actuators. Also other domains could be taken into account, but this section will focus only on the above mentioned domains, being both the most suitable and common ones. In the described approach, the model abstraction level will be maintained to a physics domain. However, in some cases, a behavioral approach could be justified by the need of simulating in a fast and less accurate way the IPEM in a system context where other complex parts must be considered and, therefore, simulated.

As already introduced, IPEM macro-models belong to discrete & power device domain. At the physical abstraction level they could interact with the following domains:

- **Analog**: the interaction with this domain is necessary because the IPEM must be controlled and analog drivers often represent the common choice. Transistor level description is a quite standard approach.

- **Digital**: the interaction with this domain is necessary because the IPEM is also supplied by digital drivers. This block will often be modeled by using a behavioral approach.
- **Sensors**: the interaction with this last domain is necessary because the IPEM should be controlled through feedback signals coming from some sensors that provide inputs and command for the driving parts.

#### 5.2.4.1 Analysis

The simulation level analyzed for the model development is mainly physical. An industrial prototype of IPEM device (courtesy of STMicroelectronics) has been used as workbench (Fig. 5.4). This module has been designed for both hybrid electrical vehicles (HEVs) and industrial applications up to 100 kW power rating. The module accommodates six power MOSFETs rated at 600 V/200 A in a three-phase bridge topology.

The objective is to generate a full coupled electrothermal model of the IPEM. The starting point is based on the elementary self-heating macro-model of the used devices. The use of macro-model suffices when the power MOSFET is in stand-alone behavior condition and it does not receive the thermal effects of other devices. In the case of a multi-chip power module, as the IPEM, the elementary self-heating macro-model becomes the first building block of a more general model that will be composed by three parts:

- Active device macro-model;
- RLC lumped network (electrical layer);
- Thermal impedance network (thermal layer).

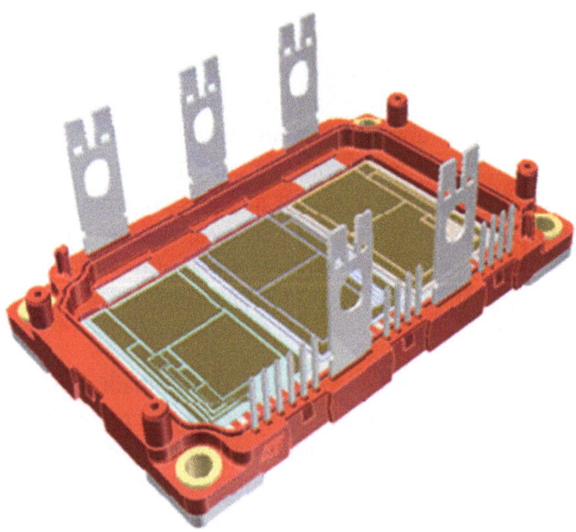

**Fig. 5.4** Schematic top view of an IPEM module with a three-phase DC–AC converter having six power MOSFETs

**Fig. 5.5** Schematic vertical stack of the IPEM

The IPEM has a common direct bonded copper (DBC) substrate and, generally speaking, it has a vertical stack as that shown, in principle, in Fig. 5.5.

If a certain power $P$ is the dissipated power within a chip, the other chips of the IPEM will undergo a junction temperature ($T_j$) elevation. In order to evaluate this rise, it is useful to know the thermal impedance curves both for the self-heating and the cross-heating behaviors. In the steady-state case, the thermal resistance $R_{\text{th}}$ is taken into account to determine the final temperature value. However, in the transient case, the thermal impedance $Z_{\text{th}}$ is the parameter that can provide information on the temperature variations. In general, a multi-die module, containing $n$ chips, is totally represented by its thermal matrix impedance shown in the following equation:

$$\begin{bmatrix} T_{j1} \\ \vdots \\ T_{jp} \end{bmatrix} = \begin{bmatrix} Z_{\text{th}11} & \cdots & Z_{\text{th}1p} \\ \vdots & \ddots & \vdots \\ Z_{\text{th}p1} & \cdots & Z_{\text{th}pp} \end{bmatrix} \begin{bmatrix} P_{d1} \\ \vdots \\ P_{dp} \end{bmatrix} + \begin{bmatrix} T_a \\ \vdots \\ T_a \end{bmatrix} \tag{5.1}$$

Since the IPEM under investigation in our case-study contains six chips, $n$ will be equal to 6. In the thermal impedance matrix the diagonal elements $Z_{\text{th}ii}$ represent the self-impedances at junction $i$ and so describe the self-heating, whereas the others elements, $Z_{\text{th}ij}$ represent the mutual impedances between junctions $i$ and $j$, hence represent the cross-heating of the devices. The matrix elements can be identified by exciting the $i$-th die by applying to it a power step and measuring the temperature response on each of the other dies. Such a procedure can be performed by using numerical simulations or experimental measurements. Once obtained the thermal impedance curves, it is possible to produce the RC networks by means of a curve fitting procedure. Our model refers to a power module with six dies and it is composed by an RC network for each die, and by the networks that connect them.

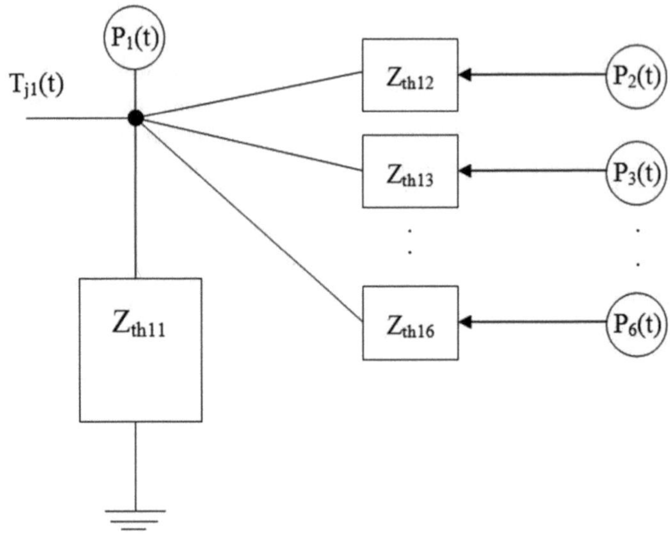


**Fig. 5.6** Block schematic of the thermal model of the device #1 (self-heating and cross-heating)

The circuital configuration of the models will be Foster-type and will have a number of RC pairs that are related to the required accuracy.

Part of the thermal model (belonging to the chip number 1) is shown in Fig. 5.6 in a block schematic. This figure shows how device 1 is affected by self-heating, by means of $Z_{th11}$, and also by cross-heating due to power dissipated into the others devices.

In Fig. 5.7 it is shown the block diagram of the electrothermal model containing all the six chips.

The final objective of this modeling activities is to produce a PSpice-like macro-model to be used for simulation purposes of an inverter application where analog driving commands and the actual load conditions are taken into account.

#### 5.2.4.2 General Data-Flow and Interfaces

The interactions of the thermal and electrical layers are depicted in Fig. 5.8, while a flow diagram related to the modeling activity under investigation is shown in Fig. 5.9. It can observed that both parts are complementary and have converged in the generation of the final macro-model of the IPEM (full thermal coupled IPEM macro-model). Co-simulation interactions have been indicated by white blocks, where several levels of interaction among simulation domains have been taken into account.

The implementation of the described flow, in general terms, may be split in the following four main phases:

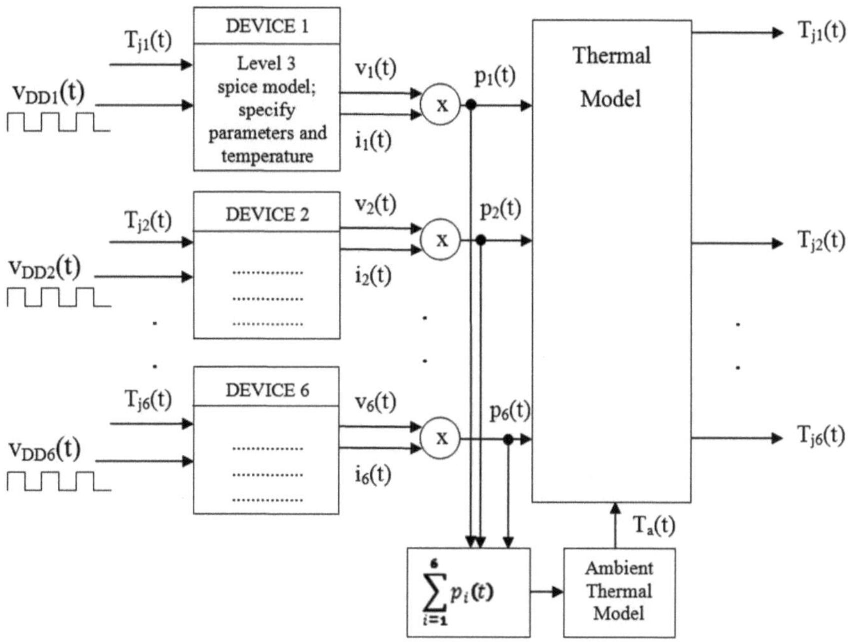


**Fig. 5.7** Block diagram of the electrothermal model of a power module

1. **CAD IPEM representation**—In this phase, the physical structure of the module has been translated in a CAD format through a 3D modeler. This action represents a key step because the produced results have been provided as input to the next phases where physical modeling issues will be analyzed.
2. **RLC lumped-parameter parasitic-network implementation**—This phase consists in implementing the RLC parasitic lumped-parameter network (electrical layer) by using the Ansys Q3D extractor software. Accurate evaluation of all relevant physical contributions is necessary in order to simplify the whole complex CAD structure and thus allowing an effective valuable simulation process.
3. **RC thermal network synthesis**—In this section of the flow diagram, starting from a series of thermal simulations performed by the thermal module of the COMSOL software package, a thermal impedance matrix has been extracted and then used for synthesizing, through a fitting methodology, the RC mutual thermal network (thermal layer) that models the thermal behavior of the module.
4. **Macro-model assembling**—In this last phase, all the relevant data retrieved in the previous phases have been merged and assembled to obtain the final macro-model. Hence, the self-heating power MOSFET macro-model has been coupled together with the thermal and the electrical layers in a feedback configuration where temperature and electrical quantities have been linked to enable a global simulation in the frame of the known constraints of different time constants.

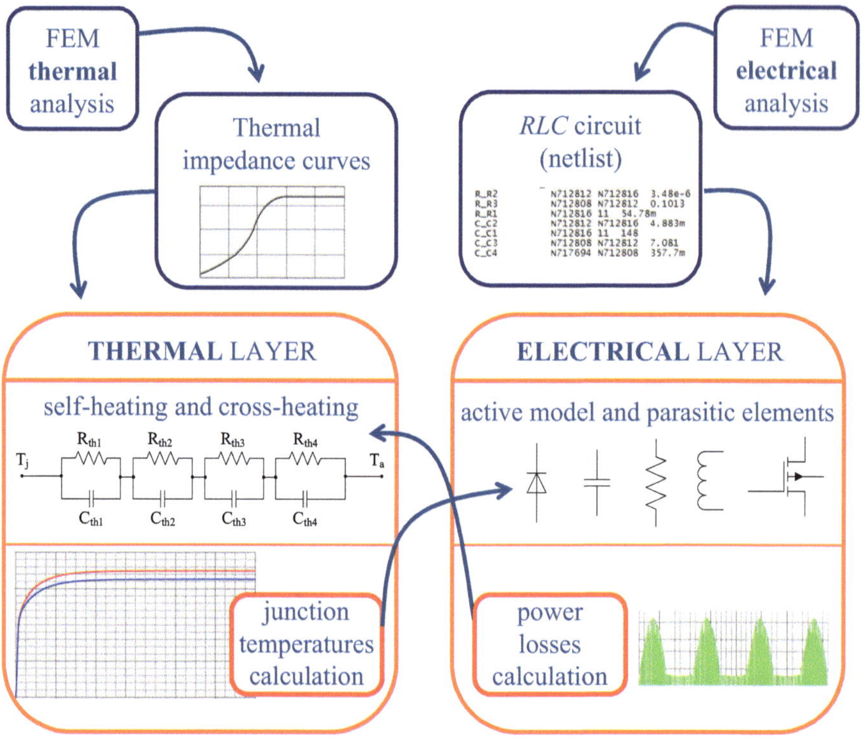


Fig. 5.8 Thermal and electrical layer interaction on IPEM

#### 5.2.4.3 Self-Heating Macro-Model of the Devices

According to the considered power module, the macro-model of each device is represented by a sub-circuit consisting of MOSFET level 3 SPICE model that is combined with controlled current and voltage sources and some passive elements, as depicted in Fig. 5.10 [17].

In the described macro-model, component G4 performs a multiplication of $V_{DS}$ (detected by (IN1+)–(IN1−) by Id) that is supplied by component H1 that transduces the current into a voltage. G4 then transduces the power into a current that is supplied to the thermal network. This macro-model represents an innovative modeling approach aimed at managing the thermal dependencies between electrical parameters and temperature, whose components derive from the device physics but that, taken individually, are modeled in a behavioral way. This model has been adopted as a base active element for the IPEM electrical and thermal modeling technique for the envisaged case-study.

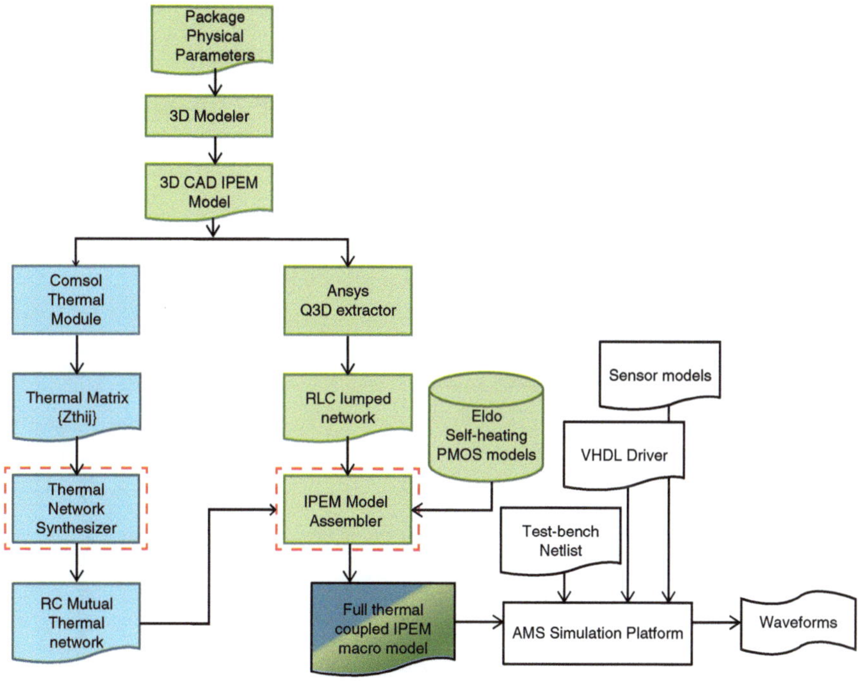


**Fig. 5.9** Thermal and electrical layer co-simulation data-flow for IPEM models

#### 5.2.4.4 Parasitic Electrical Layer Extraction

To model the physical structure of the IPEM and to perform the electromagnetic analysis aimed at extracting parasitic contributions, we take advantage of the Ansys Q3D Extractor suite. The integrated power module, designed for HEV, accommodates six power MOSFETs with an embedded current-sensing minor-MOSFET in a mirror configuration. As known, the power MOSFET features an embedded body diode. The module is enclosed in a plastic case that with its five embedded terminals provides 19 I/O pins to the module (Fig. 5.11). We are interested in extracting the parasitic inductances, resistances, and capacitances of the whole module by considering the passive contributions of the six dies.

The first step consists in drawing the IPEM geometric model; to do this action it is necessary to have a detailed geometric description of all components of the module such as package, PCB, traces, dies, ribbons, bonding wires, etc. The next step consists in assigning the material properties to the objects of the module and to the background environment. The main objects to be drawn, the related materials, and their electrical properties are described in Table 5.1.

A material is defined by its electrical properties such as the relative permittivity, relative permeability, bulk conductivity, and dielectric loss angle. The electrical properties of the materials could be retrieved either by the providers of the

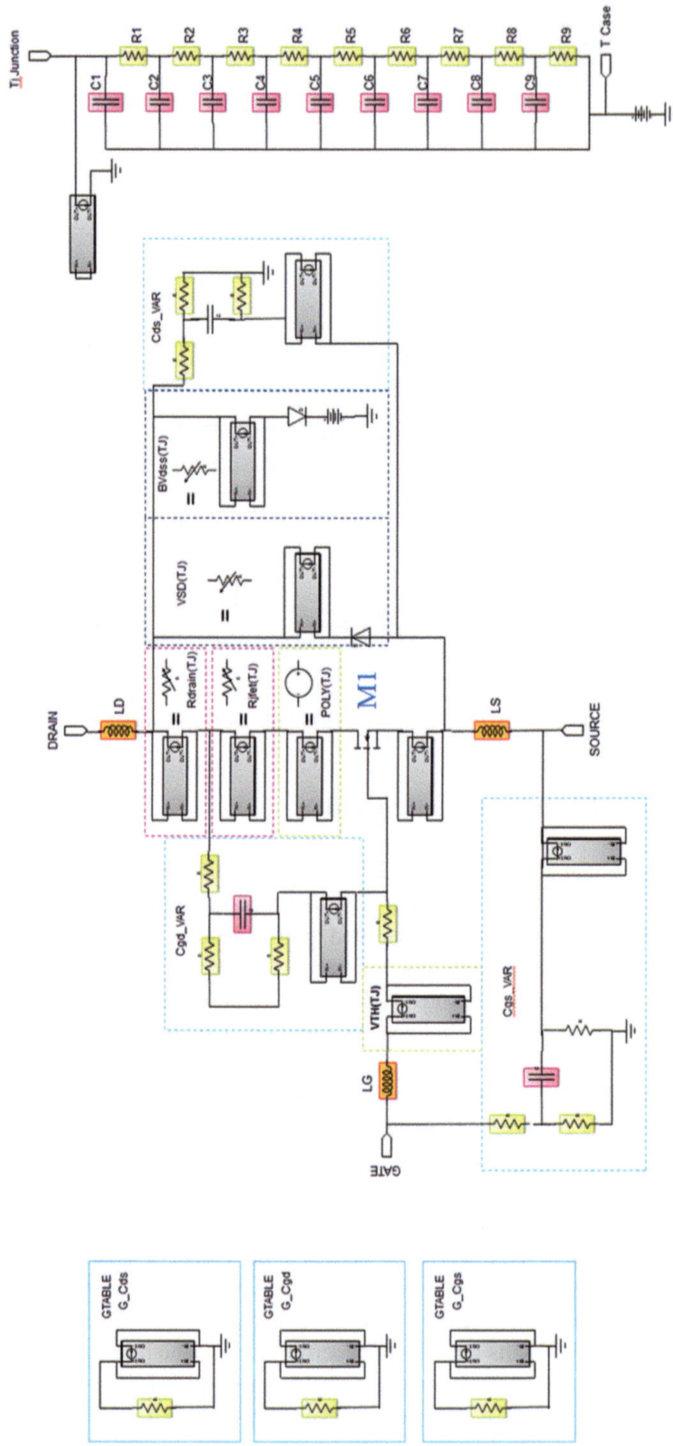


**Fig. 5.10** Elementary self-heating macro-model cell with the main components details of a power MOSFET

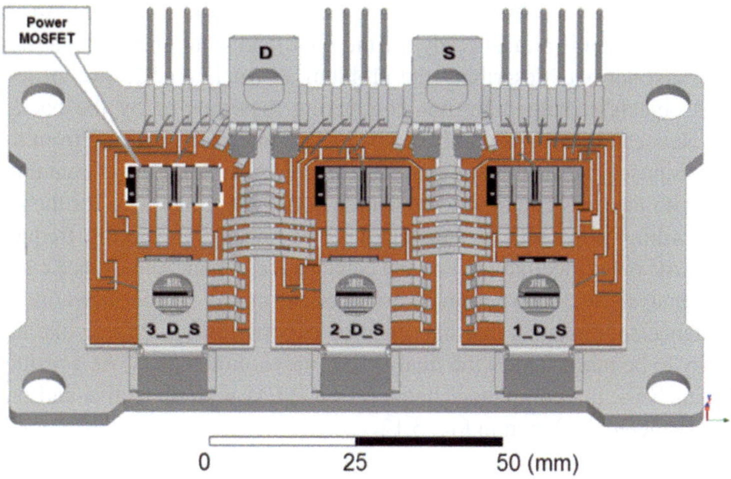


**Fig. 5.11** Internal structure of the case-study IPEM: 3_D_S, 2_D_S, and 1_D_S are the AC terminals, while S and D are the DC terminals, also represented are the control and sensing pins

**Table 5.1** Sub-component of the IPEM, materials, and their electrical properties

| Sub-component | Material | Electrical properties |
|---|---|---|
| Ribbon, Bondwire | Aluminum | Relative permittivity = 1<br>Relative permeability = 1.000021<br>Bulk conductivity = 38,000,000 S/m<br>Dielectric loss tangent = 0 |
| Pin, Trace | Copper | Relative permittivity = 1<br>Relative permeability = 0.999991<br>Bulk conductivity = 58,000,000 S/m<br>Dielectric loss tangent = 0 |
| Die | Silicon | Relative permittivity = 11.9<br>Relative permeability = 1<br>Bulk conductivity = 0 S/m<br>Dielectric loss tangent = 0 |
| DBC | Alumina | Relative permittivity = 9.2<br>Relative permeability = 1<br>Bulk conductivity = 0 S/m<br>Dielectric loss tangent = 0.008 |

components, or by the literature, or by a library of materials inside the reference manual of the package Ansys Q3D Extractor.

Once completed the drawing of all featured components and assigned the pertinent materials, it is possible to define reference simulation ports identified by the tool as "source" and "sink" terminals. By definition, the sink terminal collects all the currents while the source terminals act as current injector, once assuming that the currents enter from the die ports and leave at the end of the pins. Usually we need to put a source at the end of the die port, and a sink at the end of the pin. At this stage, the model is almost ready for the final field solution. The final step is to define the distinct nets describing the problem. A net is a collection of touching conductor objects separated by non-conducting materials or by the background material. Nets can be assigned only to conductive materials. Starting from the pins a net is composed by all touching conductor objects linking those containing the sink port with the ones containing the source ports [18]. Finally, as for the solution setup, according to problem characteristics, we specify a working frequency at 50 kHz for AC resistance/inductance and capacitance/conductance. Once completed the FEM analysis we have retrieved a numerical matrix of resistance/inductance or capacitance/conductance that may then be exported into a PSpice-like lumped-model easy to be integrated in the final IPEM macro-model [5.1]. As a result of the previous analysis, the complete scheme of the IPEM with the active and parasitic lumped parameters is shown in Fig. 5.12.

#### 5.2.4.5 Thermal-Layer Implementation

In the case of a multi-chip power module, it is important to take into account the cross-heating effect among the dies and how it elevates the junction temperature compared with the self-heating case. In the developed approach, the final objective

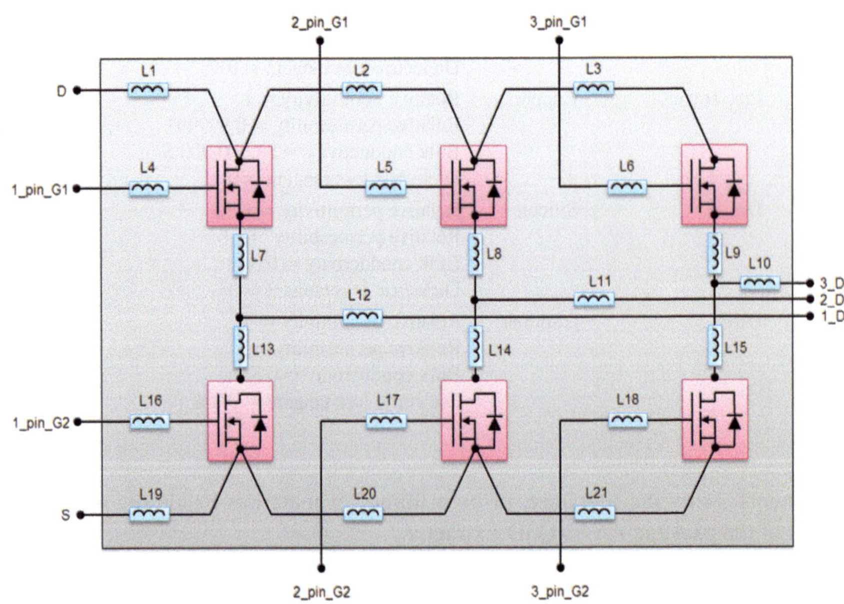


**Fig. 5.12** Equivalent circuit including part of the Q3D extractor tool output of an IPEM (only parasitic inductances) and the active part of the DC–AC three-phase power converter (inverter)

for the implementation of the thermal layer is the synthesis of an electrothermal impedance network that has been accomplished by the following four main steps:

1. FEM thermal simulations of the IPEM;
2. Thermal impedance matrix extraction;
3. Electrothermal network topology synthesis;
4. Curve fitting for parameters extraction.

While the first step basically depends on the geometries and materials of the IPEM, the remaining steps are technologically independent since they only depend on the number of featured dies and on the number of poles used for synthetizing the Foster-type RC network. This last aspect is strictly related to the expected accuracy of the curve fitting. In order to generate the thermal impedance matrix we have performed a number of simulations consisting in applying a power step on a single die and calculating the thermal responses on all the IPEM dies.

### Thermal Simulation by a FEM Package of IPEM

The simulations have been performed by using the FEM simulator COMSOL Multiphysics [19]. In Fig. 5.13 it is depicted an outcome example of the simulation. The module has been foreseen to be mounted on a heatsink.

As it turns out, in a FEM simulation, the correctness of the boundary conditions is fundamental and in the case under investigation a convective type boundary condition has been applied to a heatsink with a heat transfer coefficient of

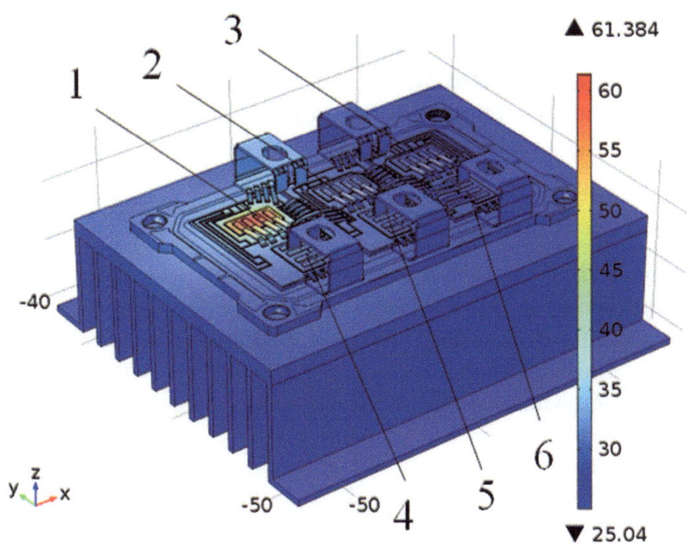


**Fig. 5.13** 3D view of a physical model of the IPEM within COMSOL after power excitation on die 1

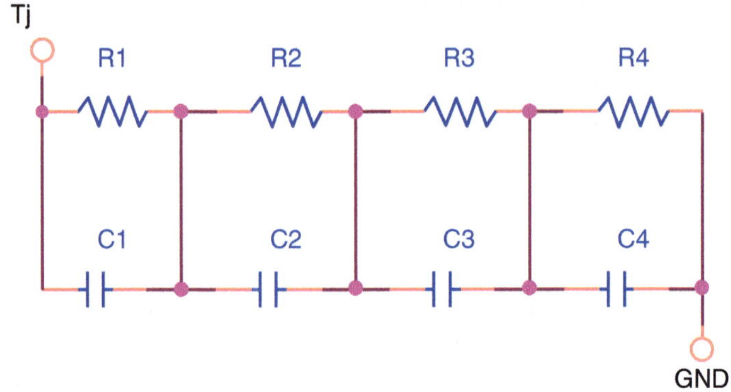


**Fig. 5.14** Example of a Foster-type network configuration with four poles

1000 W/(m$^2$K) by supposing a forced air-convection cooling mechanism. The top and lateral surfaces of the IPEM have been considered as adiabatic.

## Synthesis of the Thermal Model by a Lumped-Parameter Network

The goal is to automate the generation of the thermal-layer model by producing a spice-like Foster-type RC sub-circuit (Fig. 5.14) that has been used to merge the thermal and electrical layers together.

The curve fitting is implemented in Matlab language taking advantage of the curve fitting tool TLSynth (Thermal-Layer Synthesizer) developed for this purpose. The Matlab code is directly integrated into the tool. The requirements to use TLSynth are the installation of Matlab with the feature curve fitting tool and the Perl interpreter with Perl/TK module. The engine of TLSynth is mainly composed of two steps:

### *Thermal Impedance Curve Fitting*

In order to model the thermal-layer circuit it is necessary to fit all thermal impedance curves $Z_{th}$ retrieved by several FEM simulations of the physical model performed by the commercial software. The large amount of work required by the fitting procedure is automatized by TLSynth that takes as input the curves retrieved by COMSOL and interfaces itself with the MATLAB framework that using the Foster-type circuit equation:

$$Z_{th}(t) = \mathrm{AMP}\sum_{i=1}^{n} R_i\left(1 - e^{-t/R_iC_i}\right) \tag{5.2}$$

where AMP is an amplification factor and n is the number of Foster poles used for synthesizing the Foster-type RC network that consists of RC parallels pairs connected in series as shown in Fig. 5.14.

The fitting is done for each $Z_{ij}$ where $i \geq j$ and the outputs will be the result of the coefficients $C_k$, $R_k$ with $k = 1 \ldots n$ and the RSQUARE value that is used to measure the accuracy of the fitting.

*Circuit Model Synthesis*

The circuit model synthesis is a procedure written in Perl that, using the RC data retrieved by the fitting, generates the circuit model in Spice-like format. An outline of the whole EDA modeling flow is shown in Fig. 5.15.

A user interface (Fig. 5.16) allows the user to setup and start the synthesis. It is divided in four sections: Preferences and Activity logs, Settings data for the simulation, START button, and status bar.

In detail, the Preferences button (Fig. 5.17) opens the window that shows the general settings of the program, such as Amplification Factor, Max Step, and RSQUARE Threshold. The Amplification Factor is used to help the fitting process, the default value is 100, but it is possible to update the value in the range 1–100.

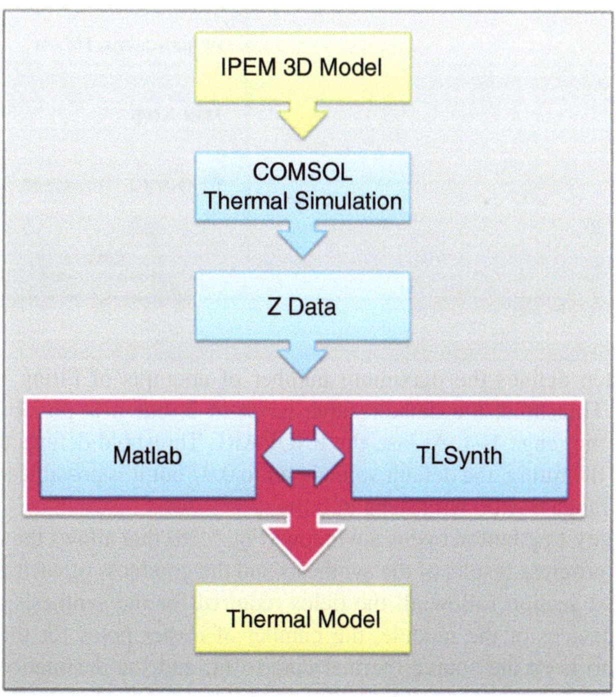


**Fig. 5.15** TLSynth flowchart

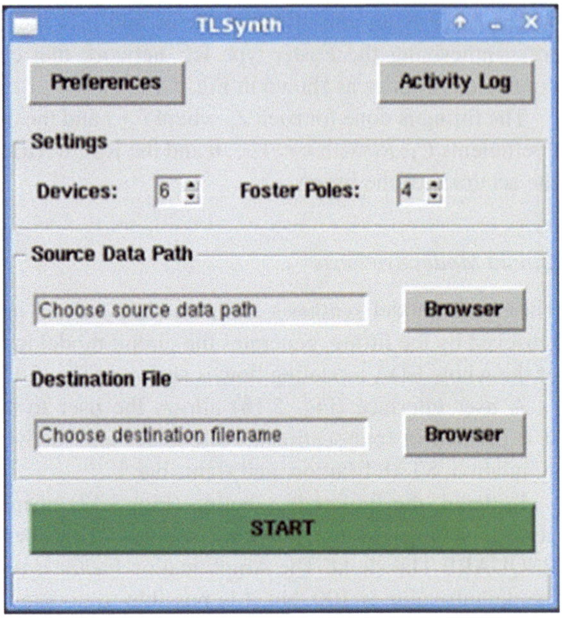


**Fig. 5.16** TLSynth graphical user interface

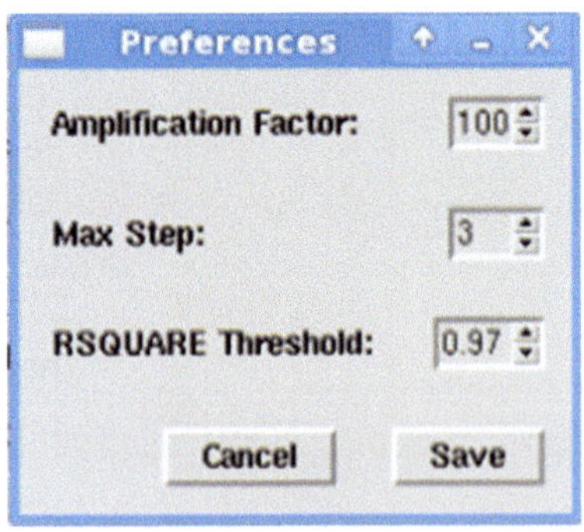


**Fig. 5.17** TLSynth preferences

The Max Step defines the maximum number of attempts of fitting to reach the RSQUARE Threshold, the default value is set to 3, but it is possible to update the value in the range 1–4. At last, the RSQUARE Threshold defines the accuracy threshold of the fitting, the default value is set to 0.97 but it is possible to update the value in the range 0–1.00 with steps of 0.01.

The Activity Log button opens a window (Fig. 5.18) that allows the user to view in detail the progress results of the synthesis and the goodness of each fitting.

In the next section follow all the fields required for the synthesis, such as the number of devices of the module, the number of Foster poles for the fitting, the path where to keep the source thermal data to fit, and the destination file where to save the synthesized circuit model in spice-like format. A default value is set to 6 for the number of devices and to 4 for the number of Foster poles. The user must compile all data to be able to start the synthesis. Once the setup is completed,

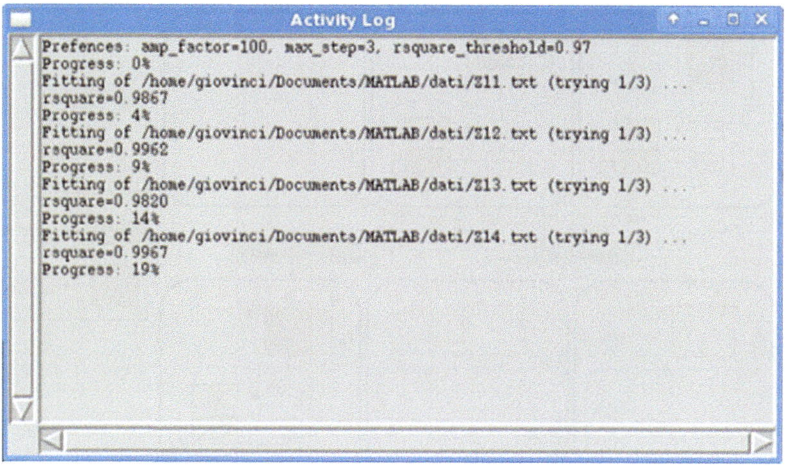


**Fig. 5.18** TLSynth activity log

the following section shows a big button in green to START the synthesis process. On the bottom of the main window a status bar is present that quickly shows the progress of the synthesis. If one of the fitting processes doesn't reach the RQUARE threshold within the maximum number of iterations set in the preferences, a warning message is shown.

Thermal Model of the IPEM

Finally, in Fig. 5.19 the thermal model of a power module containing six dies is represented. According to the above analysis the PSpice network of the thermal layer has been generated by assuming that the junction temperature of a generic die may be calculated as the sum of contributions due to the self-heating temperature, the ambient one ($T_a$), and the coupling effects due to the heat generation in other dies. The contribution due to a different die, e.g., die *2* on *1*, is evaluated by the flow of the current $I_{d2}$, that in the thermal equivalence is the power $P_{d2}$, through a current-controlled current-source (*cccs*). The $I_{d2}$ effect on the die 1, $T_{12}$, is accounted for by the specific 4-th order Foster-type network $Z_{\mathrm{th},12}(t)$ that is supplied by the current-controlled current-source.

#### 5.2.4.6 Model Validation

In order to verify the accuracy and validity of the process that has been presented to obtain a simplified thermal model, a test case has been setup as a workbench on a PSpice simulation environment. First of all a circuit schematic has been created by importing the set of generated thermal-layer netlist. Comparisons between the

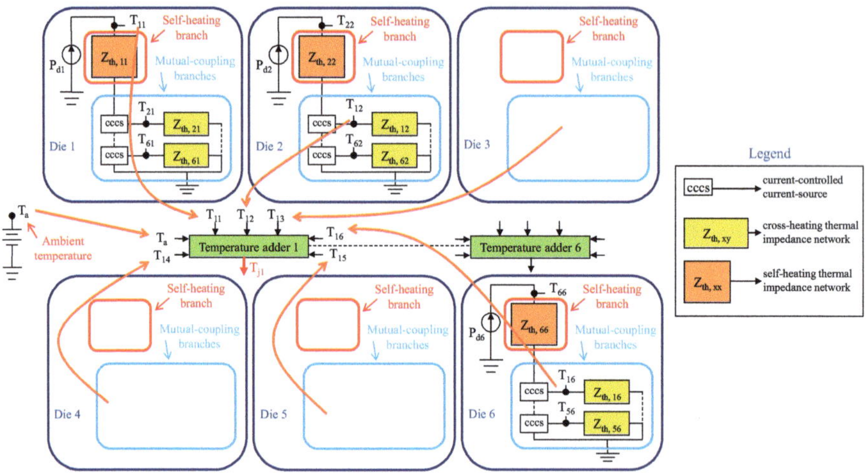


**Fig. 5.19** Block schematic of the thermal model of an IPEM

FEM data deriving from COMSOL simulations and PSpice simulated results have been done for a generic power step combination input. As a result of the comparison a good matching between the FEM and the TLSynth model waveforms was found. Figures 5.20 and 5.21 show the fine agreement between the junction temperatures on dies 1 and 4 being the simulations performed by COMSOL and by the synthesized electrothermal model. The assumption is that the dies 1-3-5 are simultaneously powered while the dies 2-4-6 are not working.

### 5.2.4.7 Summary of Features and Limitations

The proposed thermal model is suitable for both steady-state and transient analysis if the input quantities (instantaneous power losses) are time independent. In case of time dependent inputs and very small time constants of the electrical system with respect to the ones of the thermal system, average values of power losses should be used to get junction temperatures. With this assumption, we got only spot temperatures in assigned locations of the module and, in some instances, we lost detailed information in other locations. This happens because the model is not able to predict true instantaneous variations of the temperature in all the region. Moreover, in case of very fast electrical transients which have the nature of a single pulse with high specific energy production, e.g., for short-circuit conditions analysis, a more accurate thermal model at the transient level should be used to take into account the specific nature of such a phenomenon (adiabatic process).

Taking into account the limitations mentioned above, once all aspects of the FEM analysis have been defined, the PSpice-like network allows achieving the accuracy of the FEM simulation with the lower computational demand of a lumped-parameter circuit simulation.

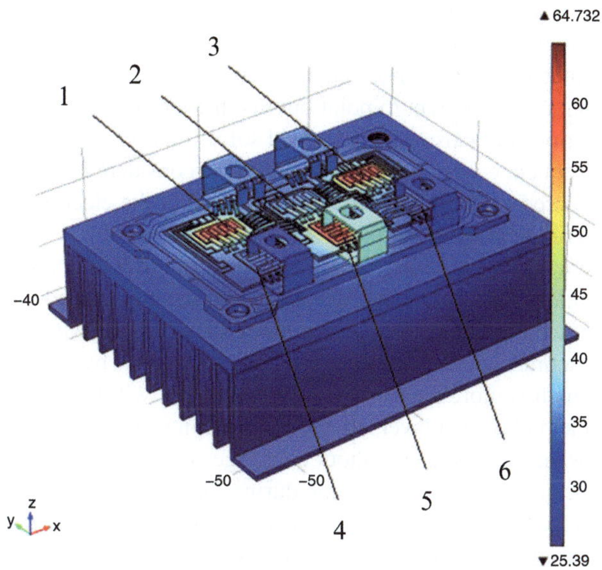


**Fig. 5.20** Surface temperatures after a time interval of 80 s that dies 1-3-5 are simultaneously powered

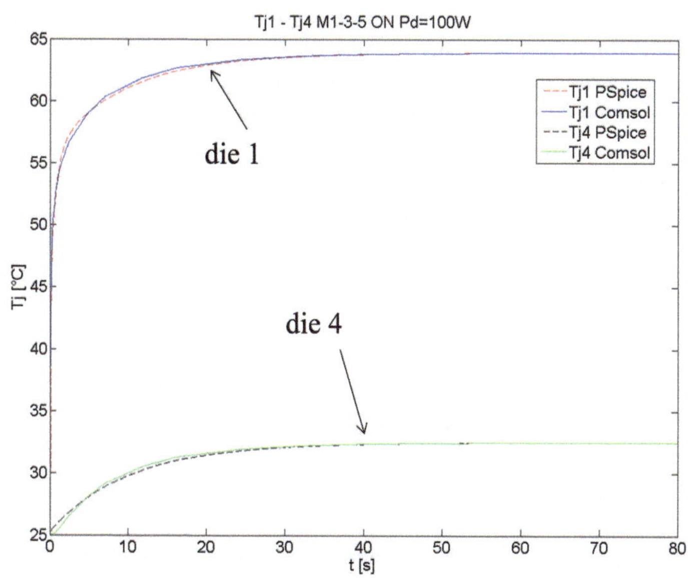


**Fig. 5.21** Comparison between the junction temperatures on dies 1 and 4 obtained by simulation runs with COMSOL and the ones retrieved by the synthetized electrothermal model by assuming die 1-3-5 simultaneously powered

Even if the physical approach is partially used, however the proposed model largely remains related to the device physics. Therefore high accuracy and validity could be guaranteed in a wide range of operations. As for the validity of the electrical characterization, it is important to point out that the tool that has been used for the parasitic extraction is a quasi-static EM field solver that poses some limitations on the range of frequency where the model could be considered valid. Usually, this model is considered applicable if we take into account the electrical length below $\lambda/10$ of the operating frequencies.

Two different approaches have to be considered in order to connect the thermal layer of the model with the active device macro-model, depending on the time constants of these layers, and specifically if they are or are not with similar order of magnitude. As a first approach we have linked the two layers by a direct connection. In this case, which is more appropriate for devices that will feature time constants of the thermal and electrical layers that are located in near orders of magnitude, or in case of transient phenomena without exchange of heat, the power losses in the devices have been evaluated in real-time during the electrical simulation and they have been directly applied to the thermal layer during a unique simulation run.

In other cases, besides the first ones already considered, the approach consists instead in the separation of the electrical and thermal simulation runs. In these cases (systems with larger time constants of the thermal layer in comparison to the electrical ones) starting from the availability of the device data sheet or the known device characteristics, and also knowing the driving algorithm of the devices and the operating conditions (current, voltage, duty cycle, switching frequency, etc.), estimations of the average power losses may be performed off line [20, 21] and such values then applied to the thermal layer. Such an approach may be also implemented by, e.g., look-up tables containing the power losses values as function of different load conditions, operating parameters, etc. The advantage of this method consists in the separation of the simulation runs having large difference of the time constant magnitudes, thus allowing short computation times and lowering the burden of data for both the systems (electrical and thermal layers) (Fig. 5.22).

### *5.2.5 Ancillary Considerations on and Applications of the Electrothermal Model*

#### 5.2.5.1 Thermal Impedance Matrix and Symmetry Issues

In general, a multi-die module containing $p$ dies is totally represented by its equations and the thermal impedance matrix (5.1), here recalled:

$$\begin{bmatrix} T_{j1} \\ \vdots \\ T_{jp} \end{bmatrix} = \begin{bmatrix} Z_{\text{th}11} & \cdots & Z_{\text{th}1p} \\ \vdots & \ddots & \vdots \\ Z_{\text{th}p1} & \cdots & Z_{\text{th}pp} \end{bmatrix} \begin{bmatrix} P_{d1} \\ \vdots \\ P_{dp} \end{bmatrix} + \begin{bmatrix} T_a \\ \vdots \\ T_a \end{bmatrix} \tag{5.3}$$

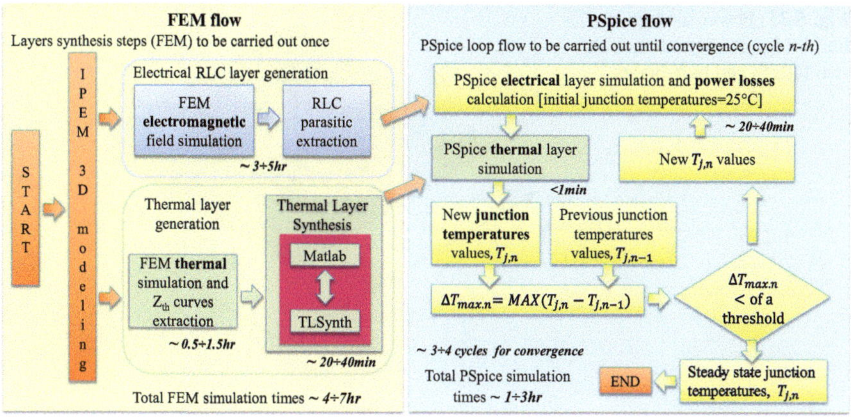


**Fig. 5.22** Layer-based electrothermal simulation approach for integrated power electronic modules

To obtain the thermal impedance matrix, a power step has to be applied to a single die and the corresponding temperature rise of all the dies have to be evaluated. Hence in the first step, die 1 is to be powered while all the dies are observed, so to retrieve the first column of the matrix of (5.3). In order to get all the elements of the matrix, this process has to be repeated for each die in the module. Thus in the worst case, the number of simulations required, or experimental measurements, is equal to that of the active dies on the module.

In the thermal impedance matrix, the diagonal elements, $Z_{\text{th},ii}$, represent the self-impedances of the junction $i$ and describe so the self-heating, whereas the others elements, $Z_{\text{th},ij}$, represent the mutual impedances between junctions $i$ and $j$, hence describing the cross-heating of the dies. In general the matrix may have a symmetric structure reflecting the physical symmetries into the module. In this case, a reduced number of different elements allow to write the whole matrix, given by the following relation:

$$k = \sum_{i=0}^{p} (p - i) \tag{5.4}$$

Furthermore, if there are identical dies on the module, the number of different elements of the matrix can be further reduced as well as that of the simulation runs or the required laboratory measurements. For the module shown in Fig. 5.23 [22], composed by two IGBTs and two diodes in half bridge configuration, the thermal impedance matrix has rank four, thus having $k = 4 \times 4 = 16$ elements, but it contains only six distinct parameters.

$$k = (4 - 1) + (4 - 2) + (4 - 3) + (4 - 4) = 6 \tag{5.5}$$

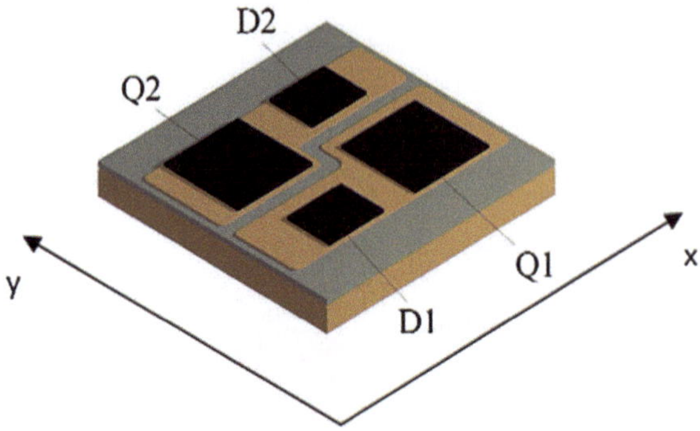


**Fig. 5.23** Power module in half bridge configuration with two IGBTs and two diodes

In fact, besides the two self-heating elements, $Z_{Q1Q1}$ and $Z_{D1D1}$, there are the cross-heating ones between the two IGBTs and between the two diodes, $Z_{Q1Q2}$ and $Z_{D1D2}$; the last two elements are due to the cross-heating in $x$ and $y$ directions, as $Z_{Q1D1}$ and $Z_{Q1D2}$. Although the distance between Q1 and D1 could be the same with respect to that between Q1 and D2, because Q1 and D1 are placed on the same layer whereas Q1 and D2 on different layers, it occurs that the cross-heating effects will be different. The only six different elements contained in the thermal impedance matrix are underlined in (5.6).

$$\begin{bmatrix} T_{jQ1} \\ T_{jD1} \\ T_{jQ2} \\ T_{jD2} \end{bmatrix} = \begin{bmatrix} \underline{Z_{Q1Q1}} & \underline{Z_{Q1D1}} & \underline{Z_{Q1Q2}} & \underline{Z_{Q1D2}} \\ Z_{Q1D1} & \underline{Z_{D1D1}} & Z_{Q1D2} & \underline{Z_{D1D2}} \\ Z_{Q1Q2} & Z_{Q1D2} & Z_{Q1Q1} & Z_{Q1D1} \\ Z_{Q1D2} & Z_{D1D2} & Z_{Q1D1} & Z_{D1D1} \end{bmatrix} \begin{bmatrix} P_{dQ1} \\ P_{dD1} \\ P_{dQ2} \\ P_{dD2} \end{bmatrix} + \begin{bmatrix} T_a \\ \vdots \\ \vdots \\ T_a \end{bmatrix} \tag{5.6}$$

#### 5.2.5.2 Boundary Conditions and Cross-Heating

In a power module, the cross-heating effects depend on the cooling mechanism. For a simple structure constituted by two identical dies placed on a copper case, as shown in Fig. 5.24, and having size of 80 mm × 50 mm × 3 mm for the case and 10 mm × 10 mm × 0.28 mm for the dies, two different kinds of boundary conditions have been taken into account. If the module is cooled by using an ideal heatsink, that is implemented by means of a boundary condition with a constant temperature (e.g., $T_c = 25$ °C) applied to the bottom of the case, the rise of the temperature of the die 2, due to a power dissipation within the die 1, is negligible because the heat flux is mainly directed in the $z$ direction, as shown in Fig. 5.24. In this case there is no lateral heat spreading in the case and the dies can be considered as in stand-alone working conditions. The hypothesis of applying a constant temperature is equivalent to using an infinite heat transfer coefficient and less than one second of simulation is required to achieve the steady-state temperature of the module. The

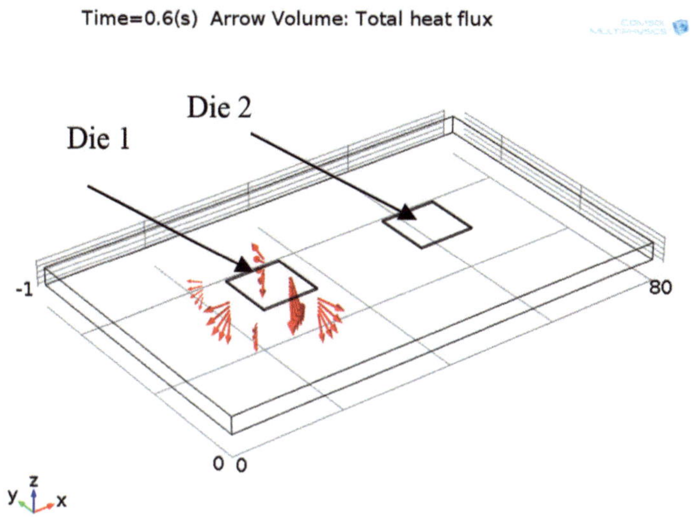


**Fig. 5.24** Heat spreading with ideal h for the heatsink ($T_c = 25$ °C)

thermal resistance matrix is represented by (5.7). Due to the symmetry, only one simulation is required to have the matrix elements.

$$[R_{th}] = \begin{bmatrix} 0.074 & 0 \\ 0 & 0.074 \end{bmatrix} \tag{5.7}$$

As a different boundary condition is created, when the module is cooled by means of a realistic heatsink, by applying a heat transfer coefficient (e.g., $h = 1000$ W/(m$^2$K)), a heat spreading occurs toward the case also into the $x$ direction, as shown in Fig. 5.25. In this case, a significant rise of temperature affects the not-powered die and the junction temperature evaluation has to take into account for the cross-heating effect. The thermal resistance matrix, related to this latter case, is represented by (5.8).

$$[R_{th}] = \begin{bmatrix} 0.516 & 0.192 \\ 0.192 & 0.516 \end{bmatrix} \tag{5.8}$$

#### 5.2.5.3 Dies Layout and Cross-Heating

In case of power module containing several dies, it is necessary to optimize some parameter like the thickness of the copper base plate, the distance between the dies, the cooling mechanism besides to consider other aspects. The right choice of the base plate thickness is very important because it works as a spreader for a power module, reducing the thermal resistance. Because this property is proportional to

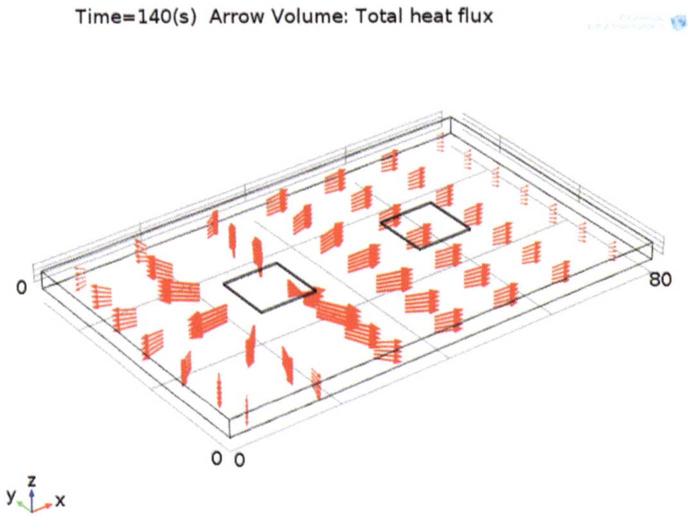


**Fig. 5.25** Heat spreading with a more common value of $h$ for the heatsink ($h = 1000$ W/(m$^2$K))

the thickness of the layer, and in particular is $R_{th} = d/\lambda A$ (where d is the thickness, $\lambda$ the thermal conductivity, and A the surface crossed by the whole power), a thinner copper layer would lead to a lower thermal resistance but only in case of one-dimensional model; conversely, in three-dimensional space, $R_{th}$ would increase due to the reduced heat spreading.

An analysis of the junction temperature variation has been performed by using the distances between the dies placed on different DBC as parameter, in order to provide some guidelines for design engineer. Simulations have been performed in order to highlight how both junction temperatures and cross-heating effects change with the distance between the dies, by keeping constant the thickness of the copper base plate, $d = 3$ mm, and by applying a heat transfer coefficient of $h = 1000$ W/(m$^2$K) as for forced air-convection cooling mechanism that exchanges with the ambient having a temperature of 25 °C.

The longer side of the dies has a length of $a = 16.4$ mm whereas the sizes of the copper base plate are 400 mm × 160 mm, large enough in order to have no impact with the dies position (Fig. 5.26).

The first FEM analysis has been performed by keeping the dies centers at a distance of $2a$ and by applying a power step of $P = 100$ W at the die 1. Other simulations have been performed by maintaining fixed the dies 2 and 5, and by moving the other dies at a distance of $3a$, $4a$, $5a$, and $6a$. The steady-state junction temperatures of die 2 ($T_{j2}$), subsequently to the simulation runs, have been written in Table 5.2 when $T_{j1}$ reached 73.25 °C at 50 s. The surface temperature, for a center die distance of $4a$, is shown in Fig. 5.27.

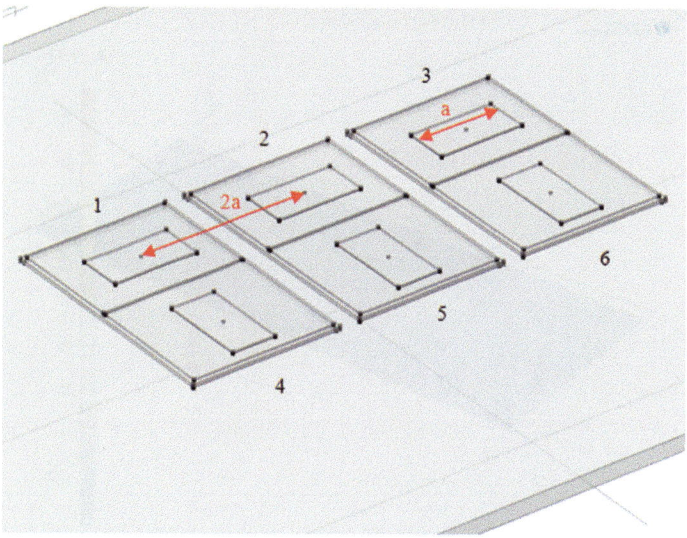


**Fig. 5.26** Geometry and dies numeration of a schematic IPEM

**Table 5.2** Junction temperatures and their variations at different distances from the heat source of device 2

| | Distance among the six dies along the $x$ direction | | | | |
|---|---|---|---|---|---|
| $n$ | 1 | 2 | 3 | 4 | 5 |
| | $2a$ | $3a$ | $4a$ | $5a$ | $6a$ |
| $T_{j2}$ (°C) | 31.44 | 28.32 | 26.79 | 25.99 | 25.56 |
| $\Delta T_{j2,n}$ according to (5.9) | 3.12 | 1.53 | 0.80 | 0.43 | |
| | $\Delta T_{j2,1}$ | $\Delta T_{j2,2}$ | $\Delta T_{j2,3}$ | $\Delta T_{j2,4}$ | |

By applying the definition of the following Eq. (5.9), the variations of the junction temperature $\Delta T_{j2}$ has been evaluated and written in Table 5.2, where $n$ refers to the five hypotheses of distance among the dies.

$$\Delta T_{j2,n} = T_{j2,(n+1)a} - T_{j2,(n+2)a} \tag{5.9}$$

Analysis of the simulation results shows that the difference of temperature is halved for each increase of the distance among the dies at step equal to the parameter a. It is easy to express such a variation by the following relation (5.10) that may be used as a rough guideline. Clearly, the above statement is tied both to the heat spreader and to the cooling mechanism.

$$\Delta T_n = \frac{\Delta T_1}{2^{n-1}} \tag{5.10}$$

The variation of $T_{j2}$ versus the distance between the dies is shown in Fig. 5.28.

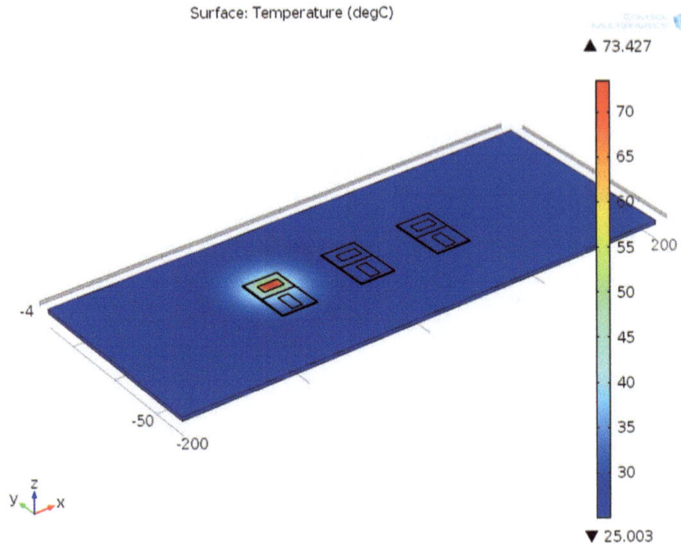


**Fig. 5.27** Surface temperature for center dies distance of $4a$

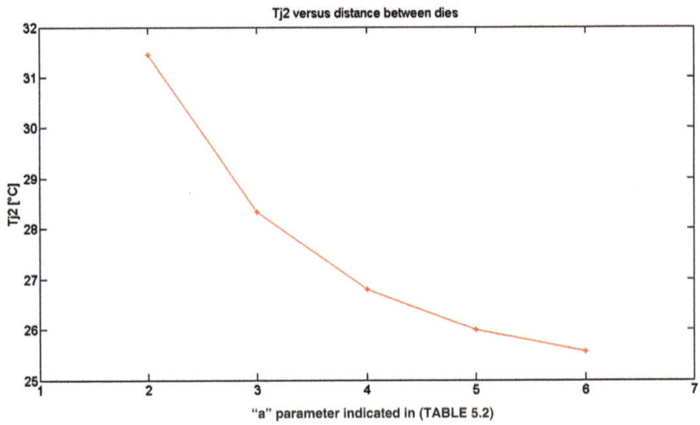


**Fig. 5.28** Junction temperature of die 2 versus the distance between the dies

When the distance reaches or overcomes $6a$, the variation $\Delta T_{j2}$ becomes lower than 2 % if compared to the previous case obtained for $5a$. Thus, it is not suitable to increase the distance. From another point of view, it is not suitable to place the dies at large distance due to the parasitic inductances, resistances, and capacitances that are associated to the module interconnections (metal path, ribbons, bond-wires). Furthermore, an increase of the distance between the dies goes against the expected size reduction of the modules too.

## 5.3 Electrothermal Simulation of Power Devices

Today's device and circuit simulators are standard tools in the development, characterization, and optimization of electronic systems and devices. However, device FEM electrothermal simulations are very time consuming and require powerful hardware equipment, particularly for complicated 3-D structures. Circuit simulations have been limited to electronic functions at a preselected temperature because the temperature dependences of the parameters of simulation models available today are taken into account at the best by changing the static global temperature.

In power electronic systems, in particular, temperature is one of the critical parameters due to the unneglectable self-heating effects and the fact that many properties of power semiconductor devices are strongly temperature dependent [23]. Power devices with high current/voltage capability and large dimension are required for high-power operation. However, self-heating induced thermal crosstalk between individual parts and components becomes serious, which degrades device performance or results in irreversible damage. Therefore, thermal management is crucially important for the viability of power devises.

Most simulators (such as ADS and Spectre) include recent advanced electrothermal models designed for devices in which the electrothermal feedback can be accounted for by a simple single-pole equivalent network [24]. However, only the self-heating effect is included and thermal coupling between devices is not taken into account. This might lead to markedly inaccurate results, particularly when the temperature of a transistor is affected by other close enough devices. Moreover, the single-pole thermal circuits have insufficient accuracy when describing the transient evolution [25].

To be able to simulate the inherent heating dynamically, introduction of an external equivalent thermal circuit, e.g., *RC* network and its interactive coupling with the electrical model, has to be implemented [26, 27]. However, this direct method usually takes into account 1-D heat flow. Assumption of a 1-D heat flow may be insufficient for power devices and large power loads in which an inhomogeneous temperature distribution caused by the 3-D heat flow from the semiconductor chip to the package and cooling assemblies can cause an inhomogeneous distribution of the electric properties of the power device along the whole chip. Building up a 3-D equivalent thermal mesh is difficult and requires structure simplification [28]. Moreover, in the case of the thermal linear network, the nonlinear behavior of semiconductor materials is not taken into account. Neglecting the nonlinear thermal properties of silicon leads to a significant error in the estimation of the transistor temperature [29].

One suitable solution is based on a relaxation method for electrothermal simulation [30, 31]. The relaxation method is based on coupling the separately solved thermal and electrical equations. The FEM is used for thermal simulation and a SPICE-like program is used for electrical simulation. The complex solution requires proper synchronization and data transfer. This modeling methodology is applied in power systems and integrated circuits to study electrothermal problems using a wide spectrum of electrical and thermal software tools (such as VHDL-AMS, FLOTHERM [32], Spectre, CircuitFire [33], HeatWave [34], GRADIENT [35], ANSYS [36], Sentaurus Device, and HSPICE [37]). The electrothermal simulation based on the relaxation method is generally used for comprehensive analysis, development, and optimization of various complex semiconductors devices. The advantages of the method are in the high speed of simulation and relative simplicity of implementation for full structure analysis.

This chapter presents electrothermal simulations of the power devices by several simulation methods: device FEM simulation, direct method with equivalent thermal 3-D *RC* network, relaxation method, and mixed-mode setup. Their features and limitations are analyzed. Due to highly time-consuming device FEM simulation, only the 2-D model has been used for the analysis. However, the neglected third dimension can cause incorrect results. The direct method with a designed equivalent thermal 3-D *RC* network provides very fast results. Nevertheless, to simplify the creation of the thermal mesh, not all parts of the package are taken into account. For fast 3-D electrothermal simulation based on the relaxation method, automated interaction is designed of Sentaurus Device and HSPICE. The complex system of the device consisting of a silicon chip, lead frame, bonding wires, package, and external cooling assemblies is used for the 3-D thermal simulation in Synopsys TCAD Sentaurus Device tool [38] using only thermal equations. The electrical properties are simulated at the circuit level in HSPICE. The automated interaction is provided by a designed Linux shell script directly supported in Synopsys TCAD Sentaurus. Another proposed electrothermal simulation is based on direct coupling between FEM thermal and circuit electrical simulation using mixed-mode setup [39–41] supported in TCAD Sentaurus environment [42]. The mixed-mode setup allows direct interconnection of FEM thermal model and electrical temperature-dependent circuit model. No synchronization and data transfers between two different tools are required as it is in relaxation method. The whole electrothermal simulation runs simultaneously. The advantages of the proposed methods are in the high speed of simulation and simplicity of implementation for complete high complexity structure analysis.

Power superjunction MOSFET under robust multipulse unclamped inductive switching (UIS) test [43] is used for verification of the used models and electrothermal simulation methods. The proposed methodology for fast 3-D electrothermal simulation is also applied to simulate a DC–DC converter under various operating conditions. Analysis of the temperature distribution of all components and their electrical properties at different operating conditions and PCB layer topology can improve the design and optimization of the device in a relative short time.

## 5.3.1 *Electrothermal Simulation of Power MOSFET*

### 5.3.1.1 Superjunction MOSFET Structure Description

The structure under investigation is the power superjunction vertical MOSFET [44, 45] in DPAK2 package. The 2-D cross section of the structure is shown in Fig. 5.29a. The trench gate MOSFET is created on top of an $n$++-type-doped substrate and $n$-type-doped epitaxial layer. The superjunction trench, created by $p$-type and $n$-layers around an air gap, provides charge balance and a high breakdown voltage for low ON-resistance. The models for different methods of electrothermal simulation are designed and described in the following sections.

### 5.3.1.2 2-D FEM Electrothermal Device Simulation

A thermodynamic physical model in Sentaurus Device tool is used for transient electrothermal simulations of the MOSFET. The standard Poisson equation, continuity equations for electrons and holes, and thermoelectric power are solved in FEM model. The 2-D FEM structure is created in Sentaurus device editor (SDE) [38]. The whole silicon chip consists of several hundred equal pillar cells. Simulation of such a large number of elements would be very time consuming. Therefore, only

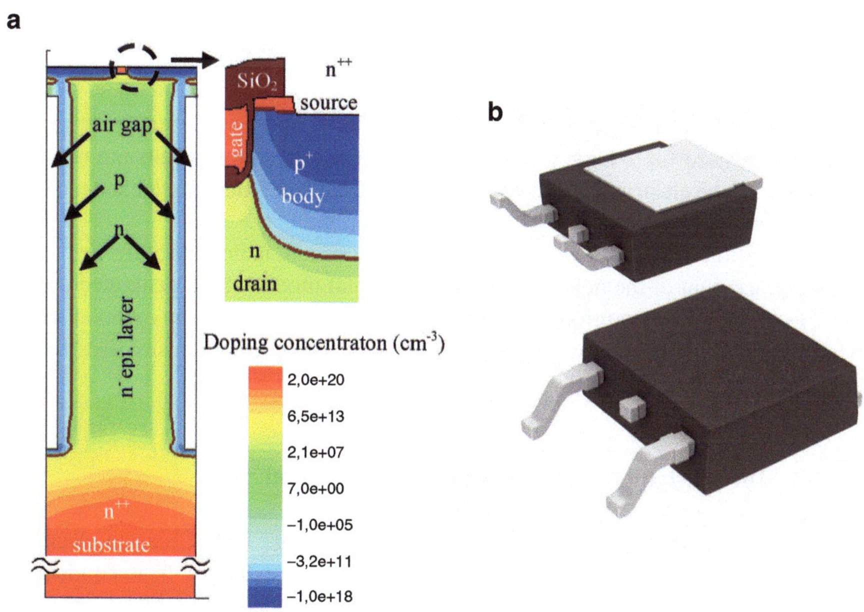


**Fig. 5.29** (**a**) 2-D cross section of the superjunction MOSFET. (**b**) DPAK2 package

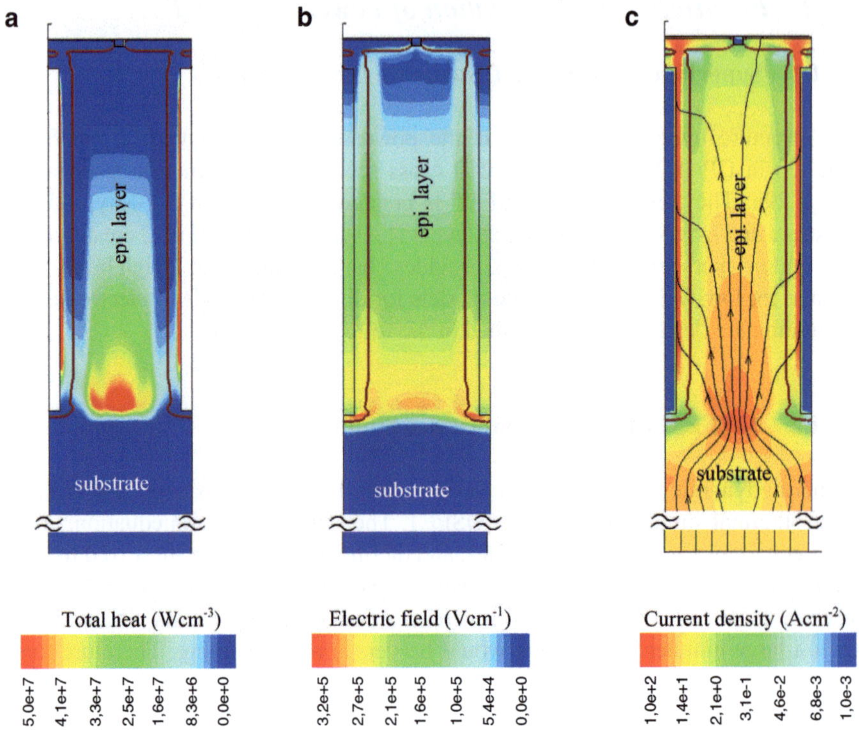


**Fig. 5.30** (**a**) Total heat, (**b**) electric field, and (**c**) current density distribution inside the structure during UIS breakdown condition

one pillar is used for 2-D device simulation. These simulation results are used for extraction of the electrical properties for the electrical analytical model and analysis of the internal behavior of the device.

Localization of the spot where heat generation occurs is important for determining the placement of the heat sources in the thermal models. Heat generated during the UIS breakdown conditions is located at the bottom of the airgap epitaxial layer, where the maximum intensity of the electric field and current density distribution are present (Fig. 5.30).

#### 5.3.1.3 Direct Method for Electrothermal Simulation

The direct method approach uses only one simulator, e.g., HSPICE as an electrical simulator. During a search for an electrical analog model for heat conduction inside structure we assume that a thermal system can be in general modeled by a discrete element electrical circuit composed of resistances and capacitances, where the temperatures and thermal powers are considered as voltages and currents, respectively (Fig. 5.31b) [26, 27]. Power dissipation of the electrical circuit is determined at all times and a current proportional to the dissipated power $I(t)$ is fed into the thermal equivalent network. The $V_{Tj}$ node voltage represents the junction temperature inside the structure. All nonlinear electrical temperature-dependent parameters of the power structure are driven by the actual temperature distribution inside the structure [46]. Due to a close relationship with physical reality of the 1-D heat flow, the parameters for the *RC* equivalent circuit diagram can be derived directly from Eqs. (5.11) and (5.12) [47]. The physical variables are specified in their thermal equivalents by using the geometries, thermal conductivities, and thermal capacities of materials, where *A* is the surface, $d_i$ is the thickness, *k* and *c* are the thermal conductivity and heat capacity of the elements. Then we can define:

$$R_i \approx R_{\mathrm{th}i} = \frac{d_i}{\kappa \cdot A} \tag{5.11}$$

$$C_i \approx C_{\mathrm{th}i} = c \cdot d_i \cdot A \tag{5.12}$$

The modification of 1-D Cauer-type equivalent thermal network to 3-D network is shown in Fig. 5.31c [48].

The 3-D thermal equivalent model of the MOSFET package is created by the *RC* network allowing 3-D heat flow. The 3-D mesh is generated automatically by Verilog-A program cycle. The thermal material properties, geometry, and dimensions are used as the input parameters for mesh creation. Only the lead frame, substrate, epitaxial layer, and package are taken into account for the simplicity of the model. Due to a strong temperature dependence of the thermal conductivity of the used materials, the thermal equivalent networks with constant elements can provide unreliable results [49, 50]. Therefore, the designed thermal network consists of temperature-dependent resistors. The resistors are driven by temperatures at relevant nodes according to a function characterizing the temperature-dependent thermal conductivity of the selected material. The schematic view of the electrical

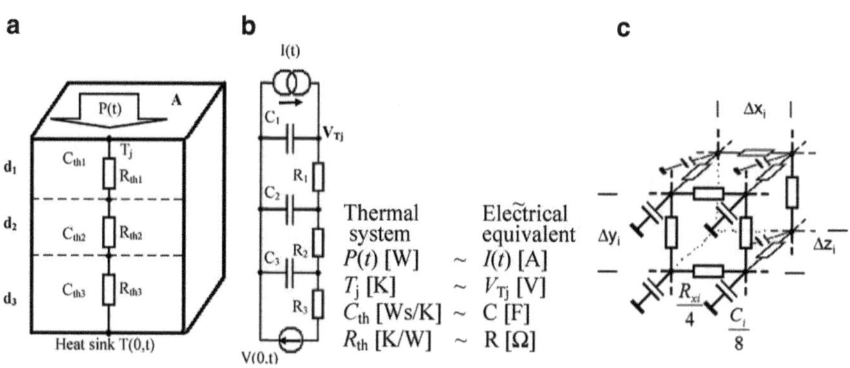


**Fig. 5.31** (**a**) Thermal structure, (**b**) Cauer electrical equivalent of the thermal system, (**c**) equivalent thermal network for 3-D heat flow

circuit diagram of 3-D equivalent thermal mesh is shown in Fig. 5.32. Referring to [51] and device simulation (Fig. 5.30), most of the heat generated during the avalanche operation of the UIS conditions is located at the bottom of the air gap epitaxial layer. Therefore, the current sources, which represent thermal heating, are placed between the epitaxial layer and the substrate.

The power MOSFET is electrically modeled with a SPICE Level 3 built in Verilog-A [52]. The electrical model is defined considering the temperature dependences of the most relevant parameters extracted from 2-D FEM device simulations of the structure at different operating temperatures. The temperature dependence of the threshold voltage $V_{T0}$, drain resistance $R_D$, body diode $D_B$, leakage current through leakage resistance $R_L$, avalanche breakdown voltage $V_{BR}$, and over current/temperature destruction SBURN are implemented in the model (Fig. 5.32) [53]. The $9 \times 9$ MOSFET segments are electrically connected taking into account parasitic resistances of the polysilicon gate electrode and metal source electrode. Figure 5.33 shows 2-D FEM simulations compared with analytical circuit simulation results of the transfer characteristics, breakdown characteristics at different temperatures, and *C–V* characteristics of the analyzed MOSFET.

#### 5.3.1.4 Relaxation Method for Electrothermal Simulation

The relaxation method for electrothermal simulation is based on the interaction of Sentaurus Device, FEM thermal engine, and HSPICE as the electrical circuit simulation program [37]. Simulation uses the relaxation method with separate but synchronized thermal and electrical simulations. Synchronization and data transfer between Sentaurus Device and HSPICE are provided by a Linux shell script. A great advantage is that the script is directly supported in Synopsys TCAD Sentaurus environment and the simulation can be easily parameterized in Sentaurus Workbench. The flowchart of the method for electrothermal simulation is shown in Fig. 5.34. After setting the initial conditions, HSPICE calculates the structure power dissipation during the first time step from the current and voltage distribution

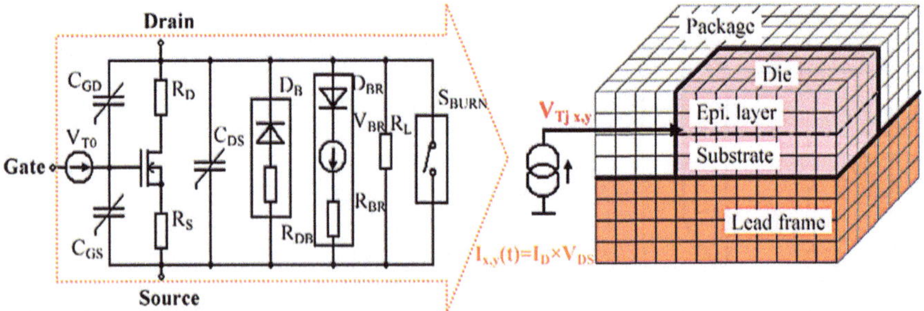


**Fig. 5.32** The electrical circuit diagram of one transistor cell (*left*) and view of the three-dimensional equivalent thermal mesh (*right*)

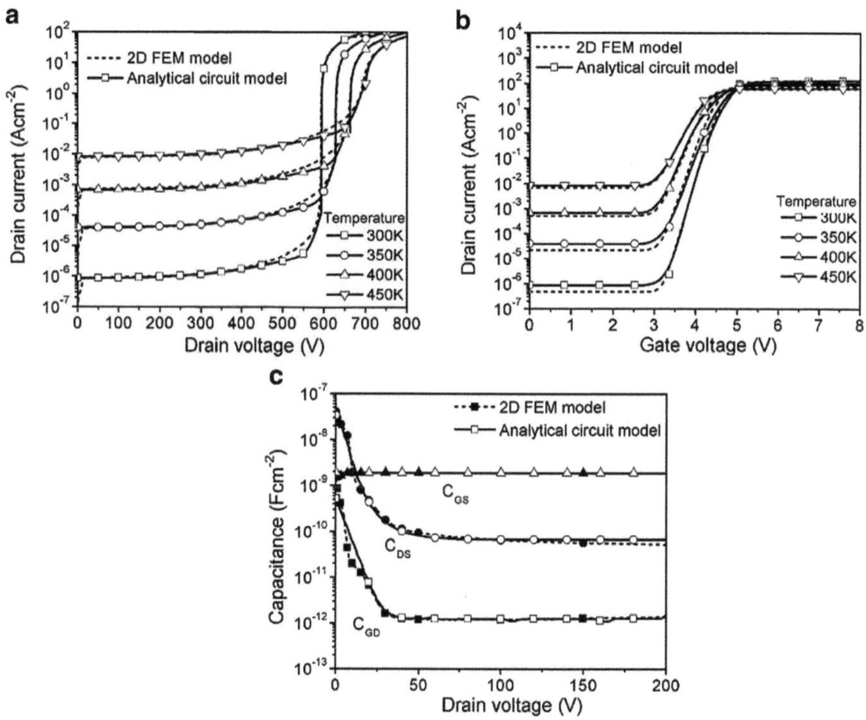


**Fig. 5.33** 2-D FEM and analytical circuit model simulation of (**a**) transfer characteristics and (**b**) breakdown characteristics at different temperatures. (**c**) *C–V* characteristics of the analyzed MOSFET

**Fig. 5.34** Flowchart of the electrothermal simulation. Linux shell script interconnects the Sentaurus Device, the thermal simulator of the structure, and HSPICE as the electrical simulation program

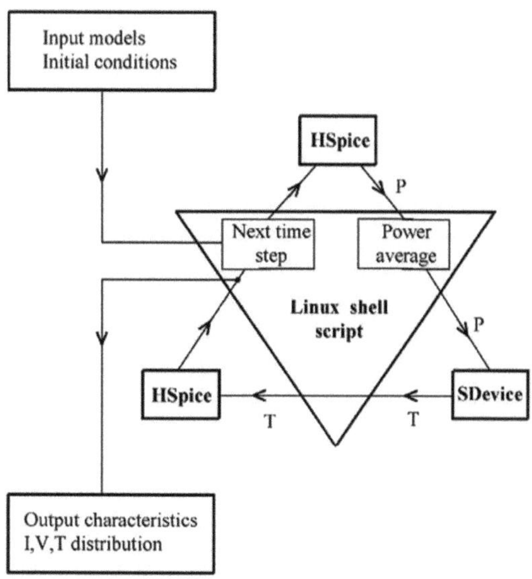

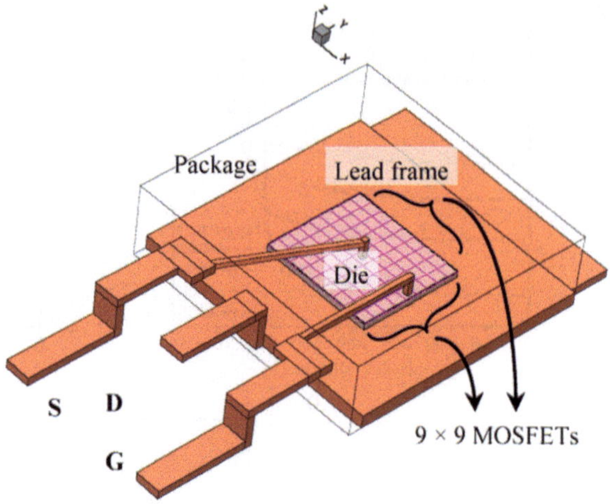


**Fig. 5.35** Thermal model of the MOSFET in DPAK 2 package with 9 × 9 thermal contacts along the silicon die

for each MOSFET region. The transient simulation is split into several time steps. The time steps are user defined and they are set to provide undistorted results for the least time-consuming computation. The average values of the powers generated during the time period between subsequent time steps are applied to the thermal contacts and Sentaurus Device calculates heat generation and heat transfer inside the structure. The temperatures in the corresponding positions of the structure are taken to drive the temperature-dependent electrical parameters of the MOSFET regions. Finally, HSPICE calculates the current and voltage distributions at relevant time step for corresponding structure temperature. The whole cycle is repeated with an increment of time until the end of simulation.

The 3-D model of the structure for thermal simulation is created in the SDE based on the physical and geometrical description of the device on chip assembled in a package with all related components (Fig. 5.35). The thermal contacts are placed uniformly on the epitaxial layer and substrate interface by a $9 \times 9$ square mesh. There occurs heat generation during avalanche conditions. The thermal contacts are generated automatically by a directly supported program cycle in the input command file, which allows simple definition of their location by the user. Each contact corresponds to one MOSFET segment and they are together interconnected by data transfer of local temperature and power dissipation. The structure boundary conditions are set to represent the heatsink to the surrounding environment. The electrical part of the MOSFET is the same as in the direct method. Sixty-four MOSFET segments are electrically connected taking into account the parasitic resistances of the polysilicon gate electrode and metal source electrode.

#### 5.3.1.5 2-D Electrothermal + 3-D Thermal Mixed-Mode Simulation

Another solution for electrothermal simulation is based on direct coupling of 2-D FEM electrothermal model of the device and 3-D thermal model of the package using a mixed-mode setup supported in Synopsys TCAD Sentaurus environment [54]. The TCAD Sentaurus mixed-mode setup can combine a multiple-device with different physical models in a circuit. The circuit netlists can contain an electrical and a thermal section. Each electrical node is associated with a voltage variable, and each thermal node is associated with a temperature variable.

The proposed solution for electrothermal simulation consists of splitting the full simulation model into a 3-D TCAD thermal model of the package (including simplified die) and a 2-D TCAD electrothermal description of one elementary cell. The two parts are connected to each other in a circuit (the so-called mixed-mode in Sentaurus Device [38]). The mixed-mode setup is built to allow heat flux exchange between the package and device via thermal node (Fig. 5.36). Drift-diffusion equations coupled to the heat equation are solved in the 2-D device model, and only the heat equation is solved in the package.

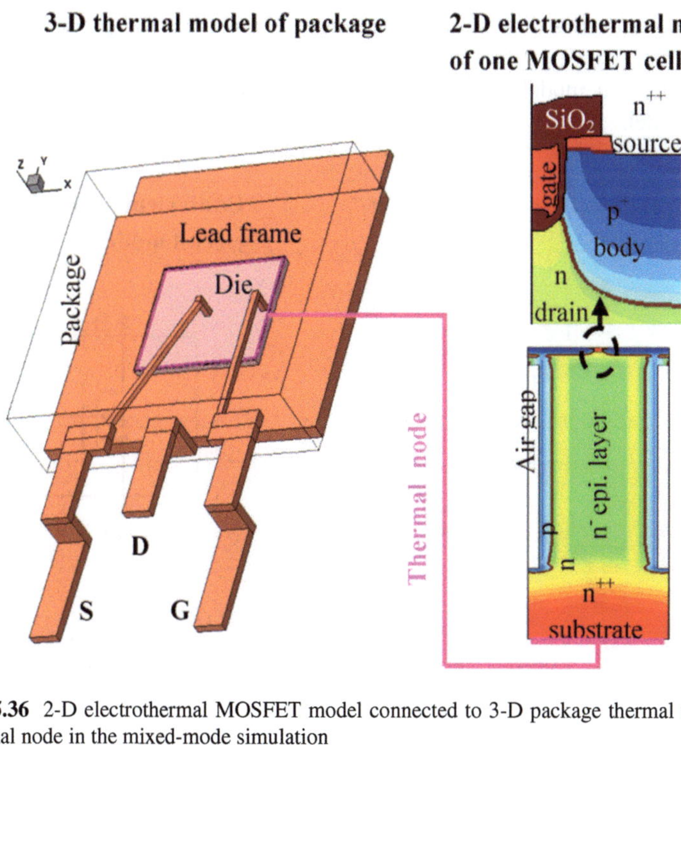


**Fig. 5.36** 2-D electrothermal MOSFET model connected to 3-D package thermal model by the thermal node in the mixed-mode simulation

#### 5.3.1.6 Electrical Circuit + 3-D Thermal Mixed-Mode Simulation

Faster simulation results can be obtained by using the equivalent temperature-dependent circuit model of the analyzed device instead of 2-D TCAD simulation [41, 54]. Sentaurus Device provides a compact model interface (CMI) for user-defined compact models. The models are implemented in C++ and are linked to Sentaurus Device at run-time. No access to the source code of Sentaurus Device is necessary. The electrical circuit model is defined considering the temperature dependences of the most relevant parameters extracted from device simulations of the structure at different operating temperatures. The temperature dependence of the threshold voltage, drain resistance, body diode, leakage current, and avalanche breakdown voltage are implemented in the model [53]. The circuit model consists of an additional power source electrode which represents heat generation inside the structure. The heat flux from this node is directly proportional to the total power dissipated in the MOSFET. This electrode is directly connected to the thermal contact of the 3-D package model through thermal node (Fig. 5.37). The temperature of the thermal node is used to drive all device temperature-dependent MOSFET properties. The fast solving of the equivalent circuit model allows splitting the silicon die into several parts and takes into account inhomogeneous distribution of temperature and electrical properties along the whole chip. The $9 \times 9$ thermal contacts are placed uniformly along the chip in the 3-D thermal model and each contact is connected with electrical circuit MOSFET model.

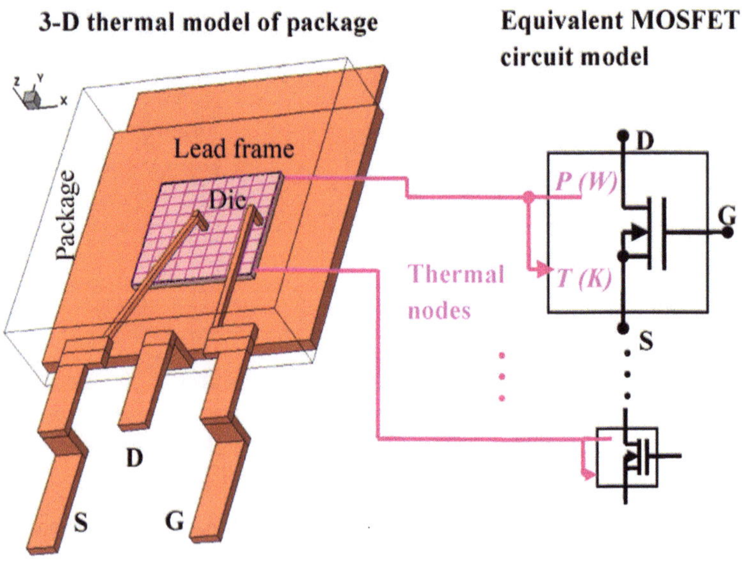


**Fig. 5.37** Equivalent MOSFET circuit model connected to 3-D package thermal model by the thermal nodes in the mixed-mode simulation

#### 5.3.1.7 Simulation Results of MOSFET Under UIS Test

The electrothermal simulation methodologies are powerful tools for analysis, optimization, and interpretation of the electrophysical behavior of power devices. The designed electrothermal models of the analyzed power MOSFET have been used to simulate the multipulse UIS test. The UIS condition represents a robust test of circuit switching operation for evaluating the ruggedness, which characterizes the device capability to handle high avalanche currents during the applied stress. The simplified UIS test circuit and corresponding current and voltage waveforms of the tested device under UIS conditions are shown in Fig. 5.38. In the case when the current flowing through the inductance is quickly turned off, the magnetic field induces a counter electromagnetic force that can build up surprisingly high potentials across the device and the whole built-in energy of the inductor is dissipated directly into the device under test [55, 56].

Figure 5.39 shows the simulation results of the multipulse UIS test using direct method with 3-D *RC* thermal network and 2-D FEM electrothermal device

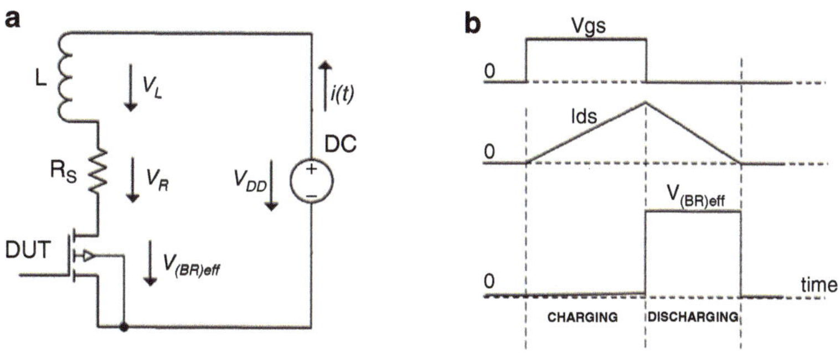


**Fig. 5.38** (**a**) Simplified UIS test circuit. (**b**) Current and voltage waveforms of the tested device under UIS test conditions

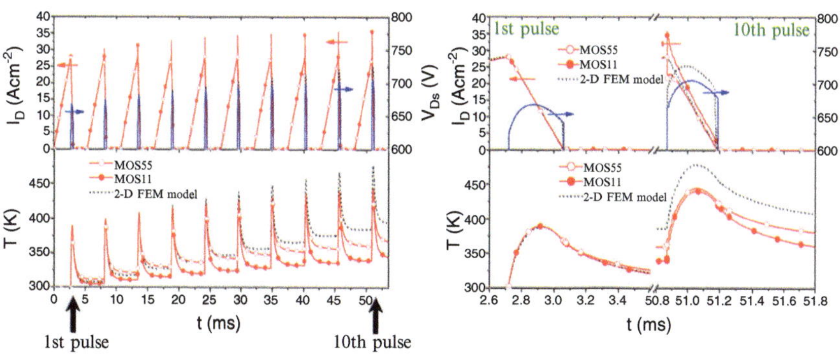


**Fig. 5.39** Comparison of the drain current, drain voltage, and temperature of the corner MOS11 and central MOS55 regions during the multipulse UIS test. Comparison of results using the direct method with 3-D thermal network and 2-D device simulation

simulation. There are comparisons of the drain currents and temperatures of the transistor segments MOS11 and MOS55, which represent the corner and center of the structure, respectively. The temperature is almost evenly distributed and the currents through MOS11 and MOS55 transistors are equal during the first pulse. After some period of the power load, the temperature and the breakdown voltage of the central MOS55 transistor cell increase more significantly in comparison with the corner MOS11 transistor cell because the corner cell is cooled more effectively. Therefore, the MOS11 current becomes higher. The 2-D device simulation considers homogeneous distributions of all parameters at third dimensions and 3-D heat flow from the semiconductor chip to the package. The inhomogeneous behavior along the whole structure is neglected. Therefore, the differences of the structure temperature and drain breakdown voltage between 3-D and 2-D simulations become more significant for the longer time simulation (e.g., the 10th UIS pulse) when the cooling of the chip by lead frame and package plays important role. The inhomogeneous distributions of the temperature and current during the 10th UIS pulse are clearly shown in Fig. 5.40.

The 3-D *RC* thermal network for the direct method is simplified and does not take into account package leads and bounding wires. The influence of bounding wires is clearly seen for the thermal model used in the relaxation and mixed-mode simulation method. Figure 5.41a shows a comparison of the drain currents $I_D$ and temperatures $T$ of the transistor segments MOS11, MOS45, and MOS55, which represent the corner, close to center, and center under bounding wire regions of the structure, respectively. The temperature and current density distributions inside the power MOSFET during the UIS test are shown in Fig. 5.41b–d. The inhomogeneities are caused by the 3-D heat flow into lead frame, package, and bounding wires. The cooling of the MOS55 segment by the bounding wire is clearly seen shortly after each pulse beginning ($t_1$, $t_{10}$). The lower temperature and breakdown voltage lead to an increase of the current density in the segment under the bounding wires.

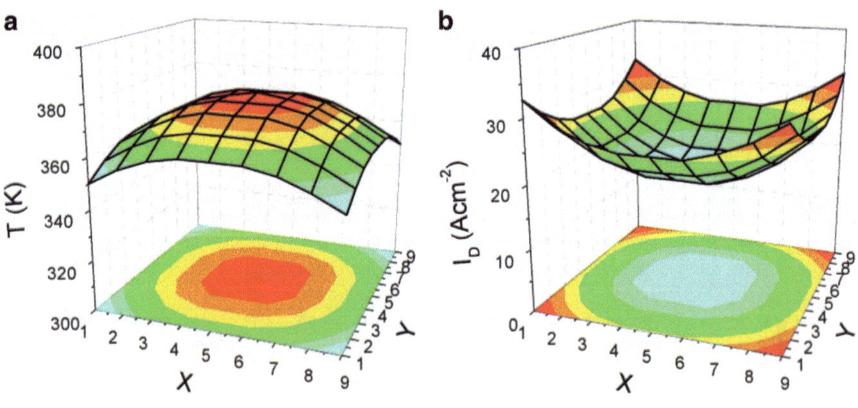


**Fig. 5.40** Temperature (**a**) and current density (**b**) distributions at the beginning of the tenth UIS pulse inside the structure

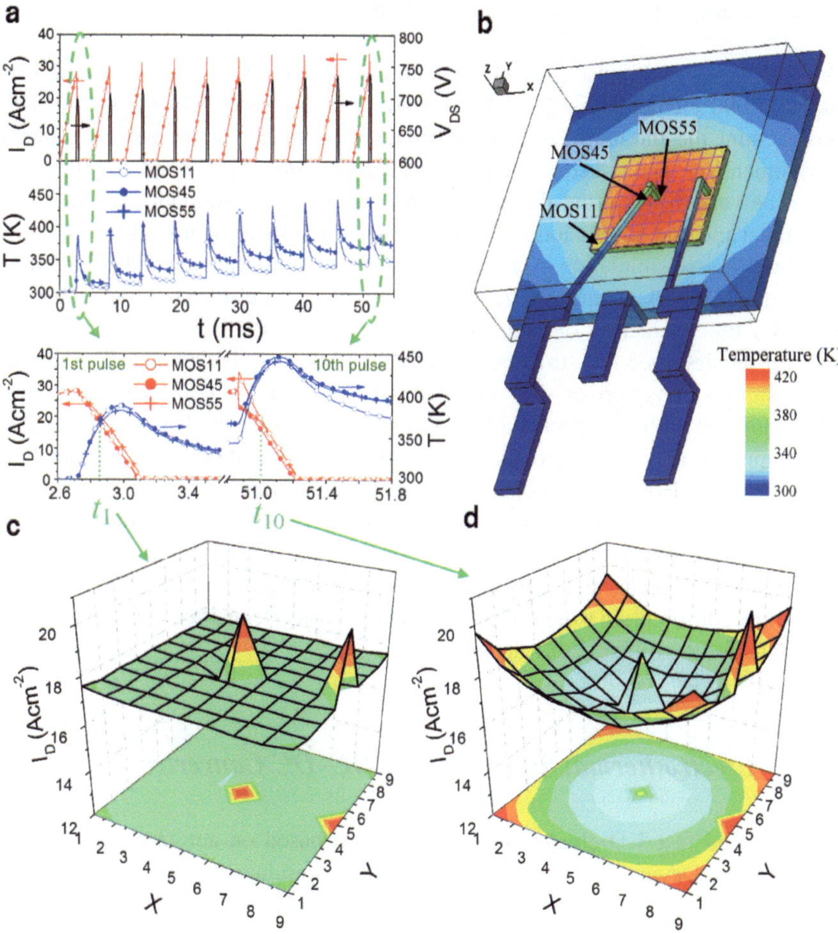


**Fig. 5.41** (**a**) Comparison of the drain current and temperature of the corner MOS11, close to center MOS45, and center under bounding wire MOS55 regions during the multipulse UIS test. (**b**) Temperature at time $t_{10}$ inside the structure. Current density distributions at time $t_1$ (**c**) and $t_{10}$ (**d**) during the UIS test inside the power MOSFET

The cooling by the lead frame and package is significant after some period of the power load (e.g., the 10th pulse). The temperature and the breakdown voltage of the close to center MOS45 segment increase more significantly in comparison with the corner MOS11 segment method. Therefore, the corner MOSFET segments current density becomes higher, as shown also using the direct method.

#### 5.3.1.8 Comparison of Simulation Methods

The 2-D FEM multipulse UIS electrothermal simulation of one pillar cell with about 17,000 elements takes $\sim 2$ h. Moreover, the 2-D simulation does not take into account the thermal flow and distributed parameters of the structure in the third dimension. The full structure FEM simulation with several hundred pillar cells would be very difficult and the simulation time would be dramatically increased. The direct method simulation of 5600 thermal network elements takes $\sim 10$ min. The short time of simulation is provided by structure simplification and reduction of the thermal network elements. The whole multipulse UIS electrothermal simulation based on the interaction of Sentaurus Device and HSPICE takes about half an hour for the designed full structure model with 17,000 mesh elements. The mixed-mode simulation setup of 2-D electrothermal model (17,000 mesh elements) interconnected with 3-D package thermal model (1000 mesh elements) takes 3 h. This approach allows full structure electrothermal simulation with thermal flow to the package. However, the distributed parameters along the whole chip are not taken into account. The whole multipulse UIS electrothermal simulation based on the mixed-mode setup of the thermal model with 17,000 mesh elements and $9 \times 9$ MOSFET electrical circuit segments takes about half an hour. The relatively lower time consumption using relaxation and mixed-mode methods represents great advantage for the full structure analysis.

### *5.3.2 Electrothermal Simulation of DC–DC Converter*

The designed fast 3-D electrothermal simulation based on automated interaction of Sentaurus Device and HSPICE is applied to simulate the DC–DC converter module in a multi-die integrated circuit (IC). Various DC–DC converter conditions are used to verify the model and electrothermal simulation method. Analysis of the temperature distribution of all components and their electrical properties at different operating conditions and PCB layer topology can improve the design and optimization of the device in a relative short time.

#### 5.3.2.1 DC–DC Converter Structure Description

The DC–DC converter is created by a high-frequency voltage regulator (HFVR) for high-current applications. The device integrates a driver IC and two power MOSFETs into an ultra-compact $6 \times 6$ mm PQFN package (Fig. 5.42). The high-side (HS) transistor is responsible for switching the inductor and the load to VIN node for a specified amount of time. The low-side (LS) transistor replaces the fly-back diode [57]. These transistors are rather large in order to ensure the lowest $R_{DSon}$ possible and therefore the lowest possible power dissipation in the system. High speed switching about 1.3 MHz with minimal power loss provides the ability

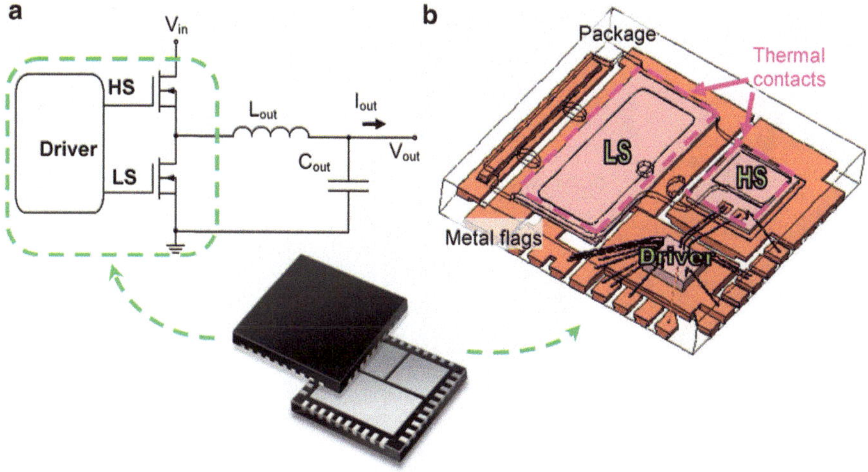


**Fig. 5.42** (**a**) Application circuit of DC–DC converter. (**b**) 3-D model of the HFVR device with included two thermal contacts for HS and LS transistor

to drive 30 A at high efficiency. The driver IC is the controlling logic along with the anti-cross protection circuitry.

The HS and LS transistors inside the multi-die HFVR IC are power vertical DMOSFET structures. The 3-D structure of the D-MOSFET is shown in Fig. 5.43a. The whole silicon chip consists of several hundred equal pillar cells. Simulation of such a large number of elements would be very time consuming. Therefore, only one pillar is used for device simulation. These simulation results have been used for the extraction of the electrical properties for an electrical analytical model and analysis of the internal behavior of the device. The power MOSFETs are electrically modeled with a SPICE Level 49 built in Verilog-A. The electrical analytical circuit model is defined considering the temperature dependences of the most relevant parameters extracted from device simulations of the structure at different operating temperatures. The temperature dependence of the threshold voltage, drain resistance, body diode, leakage current, and avalanche breakdown voltage are implemented in the model [53].

The power losses of the system are not caused only by the passive parasitic components. Considerable power losses occur during high-frequency switching. Accurate dynamic behavior of the transistor model is also important. The LS transistor needs to be turned off before the HS transistor is turned on to protect shoot-through. During the dead time, when both of the transistors are turned off, the current from inductance $L_{\mathrm{out}}$ is discharged through a forward biased body diode of the LS MOSFET. When the HS MOSFET turns on, the LS transistor body diode becomes reverse biased. Due to the reverse recovery time the diode is highly conductive until the depletion region is fully depleted. High current flow from $V_{\mathrm{IN}}$ node through HS and LS transistors causes an increase of power losses during this short time. The models of the MOSFETs have been calibrated according to the $C$–$V$

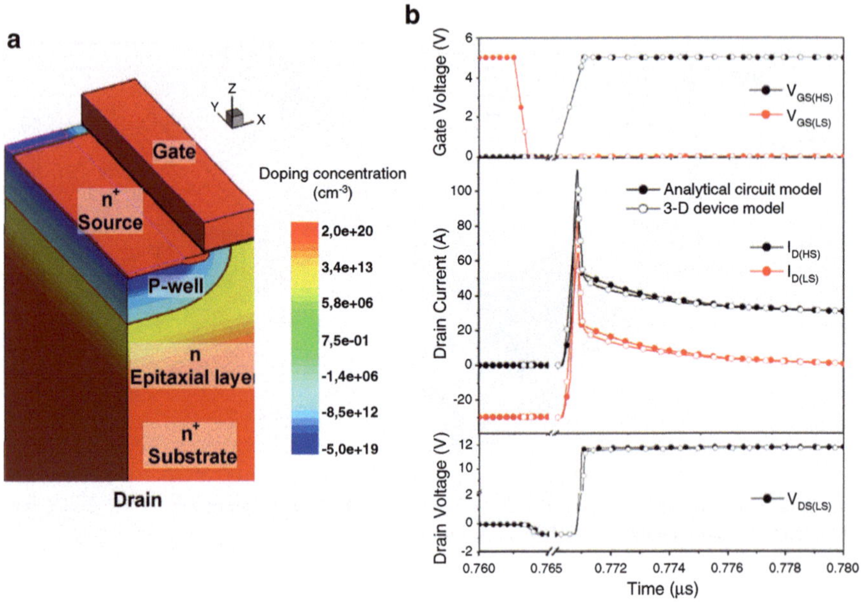


**Fig. 5.43** (**a**) 3-D structure of the D-MOSFET used for HFVR device. (**b**) 3-D device and analytical circuit model transient simulation of the MOSFETs switching

and transient characteristics to ensure correct dynamic behavior. Figure 5.43b shows device simulations of the analyzed MOSFETs compared with analytical circuit simulation results of the transient characteristics of the MOSFETs switching.

The 3-D model of the DC–DC converter for thermal FEM simulation (Fig. 5.42b) is created in the SDE. The model consists of the multi-die HFVR IC and passive components (inductor, capacitors, and resistors) placed in PCB. The multi-die IC includes two thermal contacts, which correspond to the HS and LS MOSFETs. They are interconnected by data transfer of local temperature and power dissipation with electrical circuit model. The structure boundary conditions are set to represent the heatsink to the surrounding environment.

#### 5.3.2.2 Methodology for Electrothermal Simulation of DC–DC Converter

The 3-D FEM electrothermal simulation of the DC–DC converter would be very complicated and time consuming due to the large number of pillar cells of the MOSFET, complicated thermal coupling of the silicon chips, and thermal flow into the complex system of multi-die IC and PCB. The electrothermal circuit simulation should contain an equivalent thermal *RC* network. The designed 3-D equivalent thermal mesh of the whole system from silicon chips up to PCB would be complicated or would require structure simplifications. Moreover, the temperature of the system is stabilized after a few hundred seconds. Simulation of a huge number of pulses for 1.3 MHz clock switching takes long time.

The designed electrothermal simulation methodology is based on automated interaction of Sentaurus Device, FEM thermal engine, and HSPICE as the electrical circuit simulation program. Simulation uses separate but synchronized thermal and electrical simulations. The simulation flow of the method is shown in Fig. 5.44. The transient simulation is split into several time steps. The time steps are user defined and they are set to provide undistorted results for the least time-consuming computation. Short time periods are selected for electrical simulation and extraction of power dissipations during each simulation step. The calculated powers of the MOSFETs are applied to the thermal contacts and Sentaurus Device calculates heat generation and heat transfer inside the structure. Corresponding temperatures of individual components are taken to drive the temperature-dependent electrical parameters during next time step [58]. Every other simulation step starts from the last saved point in the previous step. Sentaurus Device directly supports save and load commands. HSPICE allows save and load by initial conditions. The whole cycle is repeated with an increment of time step until the end of simulation. The reduced number of simulated pulses in HSPICE and optimized mesh for FEM thermal simulation significantly accelerates the whole electrothermal simulation.

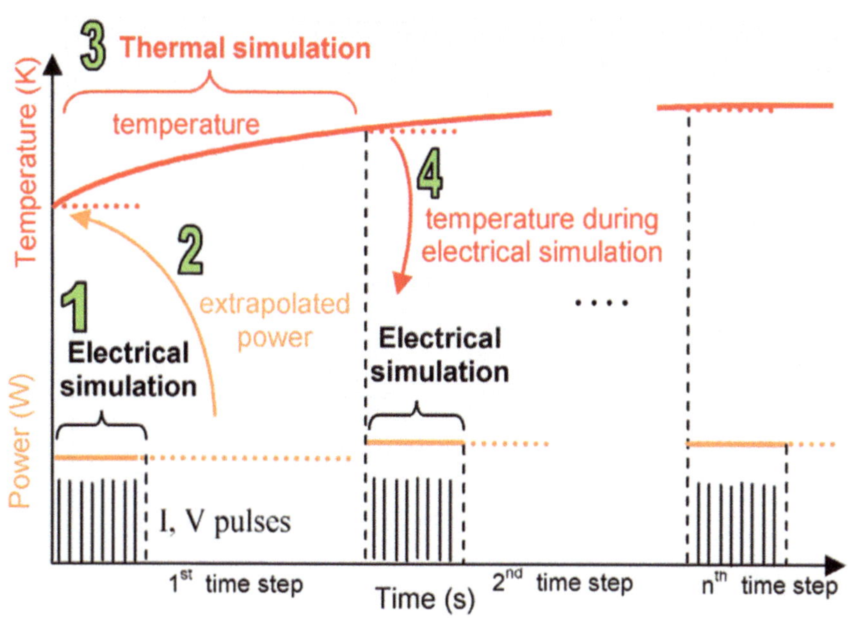


**Fig. 5.44** Electrothermal simulation flow. Short time periods are selected for electrical simulation (*1*) and extraction of power dissipations (*2*). Sentaurus Device calculates heat generation (*3*) and corresponding temperatures are taken to drive the temperature-dependent electrical parameters (*4*)

#### 5.3.2.3 Simulation Results of DC–DC Converter

The designed methodology of electrothermal simulation is used to analyze DC–DC converter under different operation conditions and PCB layer topology. The evaluations of the device component temperatures, losses, and DC–DC conversion efficiency for two different input voltages (12 V and 3 V) are shown in Fig. 5.45. According to the different operating conditions and the losses of the MOSFETs, different temperature distributions of the MOSFETs and driver chips are clearly seen in Fig. 5.46. Decreasing $V_{in}$ changes on–off ratio of both MOSFETs and increases power losses of HS MOSFET which leads to increase in its temperature. Visualization of the inhomogeneous temperature distribution can help to identify critical regions of the analyzed structure.

The differences of electrical and thermal properties for various PCB layer topologies are shown in Fig. 5.47. The designed double side PCB improves cooling of the system compared to single side PCB. The lower temperature of the device provides a higher efficiency and higher load of the DC–DC converter. The designed electrothermal simulation methodology allows easily to analyze the temperature distribution of all components placed on the PCB and their electrical properties at

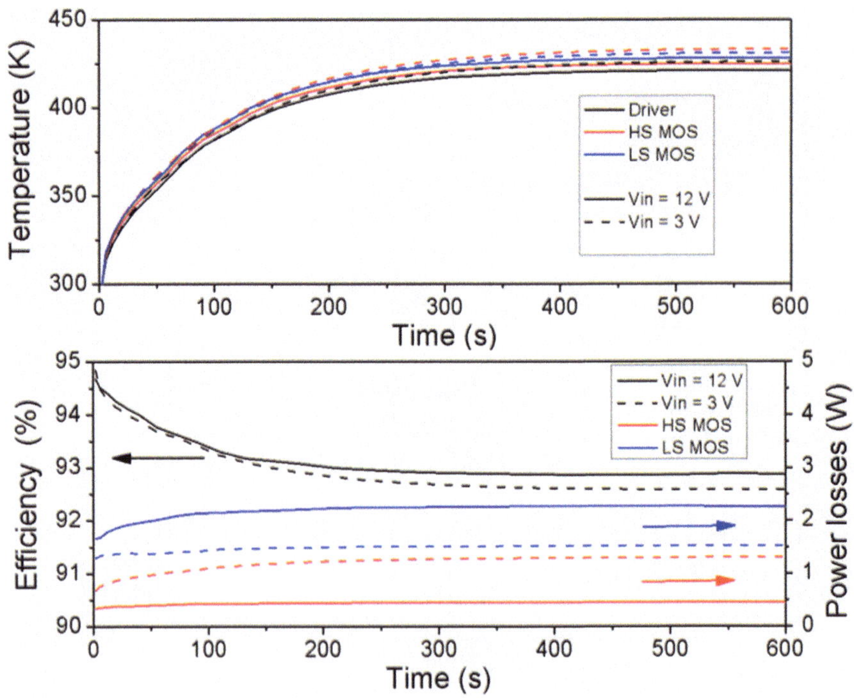


**Fig. 5.45** Evaluations of the device component temperatures (*top*), losses, and DC–DC conversion efficiency (*bottom*) for two different input voltages

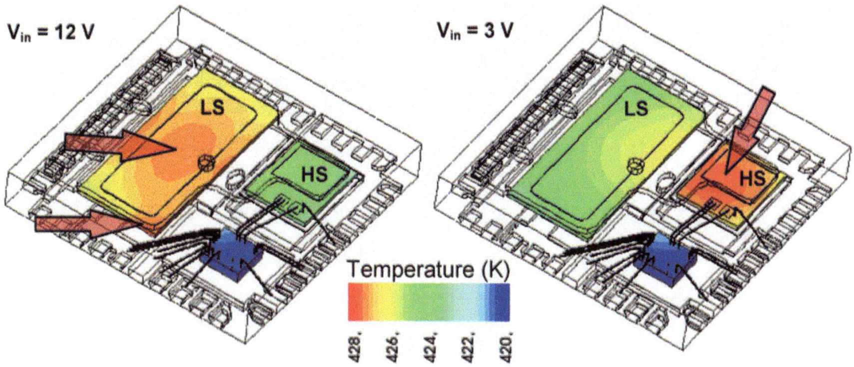


**Fig. 5.46** Temperature distributions of the MOSFETs and driver chips for two different input voltages

different conditions. Results of potential optimization of the device are provided in a relatively short time.

The use of only single 3-D FEM electrothermal simulation of one switching cycle takes about 10 min. Moreover, the thermal flow between the components is not taken into account. Circuit simulation of 1.3 MHz clock switching takes ~ 10 h till the system temperature stabilizes (600 s). The designed electrothermal simulation based on synchronized interaction of Sentaurus Device and HSPICE takes ~ 1 h. The significantly lower time consumption using the designed methodology is of great advantage for the full structure analysis and design.

### 5.3.3 Conclusion

Fast 3-D electrothermal simulations based on the relaxation method and mixed-mode setup were presented and compared with the device FEM simulation and direct method with an equivalent thermal 3-D *RC* network. The designed methodologies are developed for reducing the simulation time for complicated 3-D structures. To verify the electrothermal models and proposed methods in a harsh environment, the superjunction vertical MOSFET and DC–DC converter module were used. The simulation approaches help to assess the device robustness by means of evaluating both temperature and current distributions in the power MOSFET structures operating under critical conditions. The implemented 3-D thermal flow and distributed parameters of the structures provide more realistic simulation results. The advantages of the proposed methods are the relative simplicity of implementation, the increased speed of simulation, and the capability of full analysis of complex structures. With reference to the Section 5.3, the authors are grateful to ON Semiconductor Belgium for successful collaboration and providing information about progressive technologies and devices.

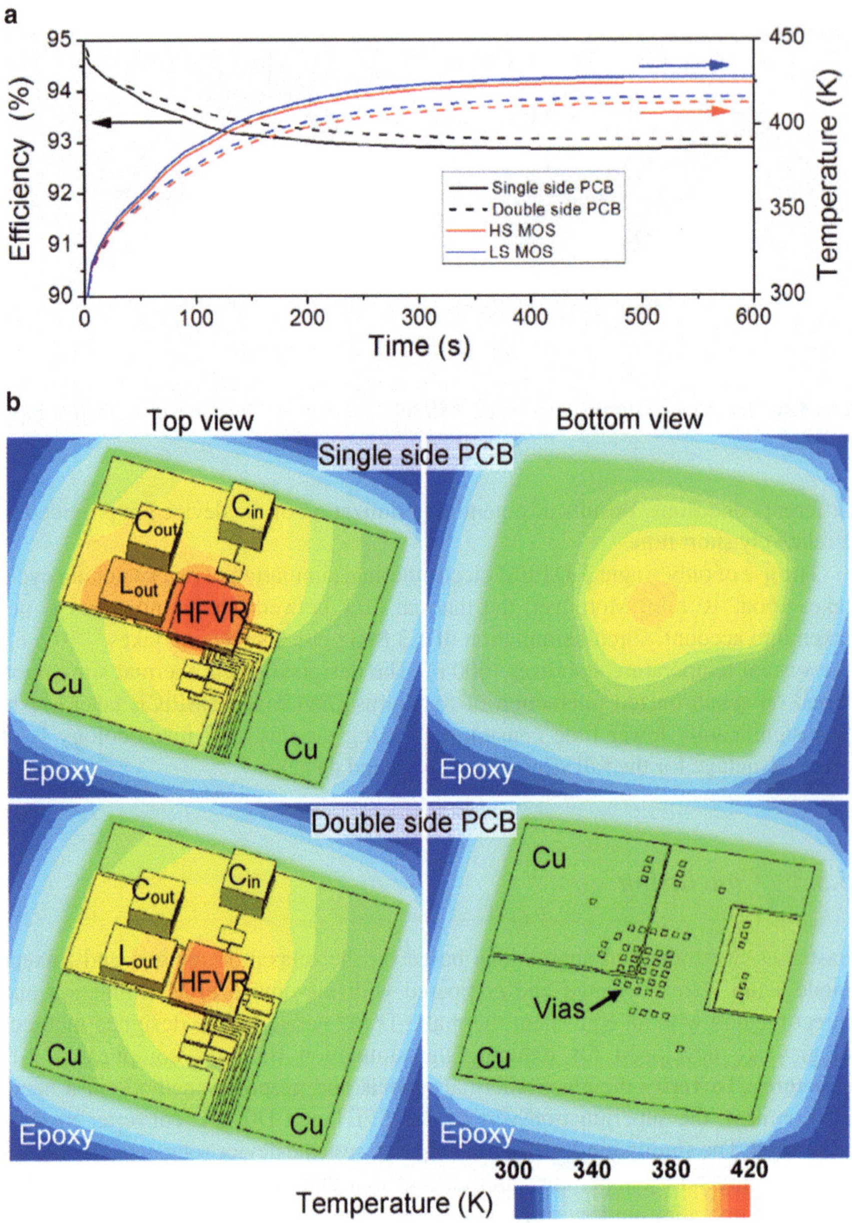


**Fig. 5.47** (**a**) Evaluations of the device component temperatures and DC–DC conversion efficiency, and (**b**) temperature distribution of the components for different PCB layer topology

## References

1. G. Bazzano, D. Cristaldi, G. Greco, A. Raciti, G. Vinci, Electro-thermal model of integrated power electronics modules based on an innovative layered approach, in *Proceedings of the 39th Industrial Electronics Society Conference (IECON)*, Vienna, Austria, 10–13 November 2013, pp. 712–717
2. D. Cristaldi, G. Greco, A. Raciti, G. Vinci, Generation of electro-thermal models of integrated power electronics modules using a novel synthesis technique, in *Proceedings of the 19th International Workshop on Thermal Investigations of ICs and Systems (THERMINIC)*, Berlin, Germany, 25–27 September 2013, pp. 216–221
3. A. Raciti, D. Cristaldi, Thermal modeling of integrated power electronic modules by a lumped-parameter circuit approach, in *Proceedings of AEIT*, Palermo, Italy, 3–5 October 2013
4. R.W. Lewis, K. Morgan, H.R. Thomas, K.N. Seetharamu, *The Finite Element Method in Heat Transfer Analysis* (Wiley, Hoboken, New Jersey, Stati Uniti 1996)
5. Z. Wan, L. Xu, Y. Zhang, X. Luo, M. Chen, J. Chen, S. Liu, Thermal analysis and improvement of high power electronic packages, in *Proceedings of ICEPT-HDP*, Shanghai, China, 2011, pp. 1–5
6. W. Zhang, D. Liu, R. Wang, X. Wang, Novel thermal analysis platform for IPEM, in *Proceedings of the 3rd International Conference on Computer and Electrical* Engineering *(ICCEE)*, Singapore, China, 2010
7. I. Guven, C.L. Chan, E. Madenci, Transient two-dimensional thermal analysis of electronic packages by the boundary element method. IEEE Trans. Adv. Packag. **22**, 476–486 (1999)
8. C.A. Brebbia, L.C. Wrobel, *Boundary Element Methods in Heat Transfer* (Computational Mechanics Publications and Elsevier Applied Science, Amsterdam, NL 1992)
9. J.Z. Chen, Y. Wu, D. Borojevich, J.H. Bohn, Bradley, Integrated electrical and thermal modeling and analysis of IPEMs, in *COMPEL*, Blacksburg, Virginia, 2000, pp. 24–27
10. Z. Khatir, S. Carubelli, F. Lecoq, Real-time computation of thermal constraints in multichip power electronic devices. IEEE Trans. Compon. Packag. Technol. **27**(2), 337–344 (2004)
11. A. Poppe, Y. Zhang, J. Wilson, G. Farkas, P. Szabó, J. Parry, Thermal measurement and modeling of multi-die packages. IEEE Trans. Compon. Packag. Technol. **32**(2), 484–492 (2009)
12. J. Reichl, J.M. Ortiz-Rodriguez, A. Hefner, J.-S. Lai, 3D Thermal component model for electro- thermal analysis of multichip power modules with experimental validation. *IEEE Trans. Power Electron*. (2014). doi:10.1109/TPEL.2014.2338278
13. P.L. Evans, A. Castellazzi, C.M. Johnson, Automated fast extractions of compact thermal models for power electronic modules. IEEE Trans. Power Electron. **28**(10), 4791–4802 (2013)
14. R.B.B. Ovando, F.A. Ramírez, C. Hernandez, M.A. Arjona, A 2D finite element thermal model of a three-phase-inverter heat sink, in *Proceedings of CERMA*, Morelos, Mexico, 2010, pp. 696–701
15. A. Raciti, D. Cristaldi, G. Bazzano, G. Greco, G. Vinci, Layered electro-thermal model of high-end integrated power electronics modules with IGBTs, in *Proceedings of the 40th Industrial Electronics Society Conference (IECON), IECON*, Dallas, TX, USA, October 29–November 1, 2014
16. A. Raciti, D. Cristaldi, G. Bazzano, G. Greco, G. Vinci, Integrated power electronics modules: electro-thermal modeling flow and stress conditions overview, in *Proceedings of AEIT*, Trieste, Italy, 18–19 September 2014
17. G. Bazzano, D.G. Cavallaro, G. Greco, A. Grimaldi, S. Rinaudo, 2D Thermal propagation analysis of discrete power devices based on an innovative distributed model technique and CAD framework, in *IEEE 16th THERMINIC*, Barcelona (Spain), 2010
18. ANSOFT, Ansys Q3D Extractor Online Help Ver. 10, 2010
19. COMSOL Inc., COMSOL Multiphysics Ver. 4.3, 2010

20. G. Belvedere, C. Guastella, M. Melito, S. Musumeci, A. Raciti, Low-voltage MOSFET with small on-resistance: an extended characterization in high-efficiency power converter applications, in *IEEE Industry Applications Conference*, Chicago, IL, USA 2001
21. G. Belvedere, M. Candelargiu, C. Guastella, M. Melito, S. Musumeci, Design considerations on the low-voltage MOSFET applications to wheelchair drive systems, in *IEEE International Symposium on Industrial Electronics*, L'Aquila, Italy 2002
22. P. Evans, C. Johnson, Fast extraction of dynamic thermal impedance for multi-chip power modules, in *Proceedings of International Conference on Integrated Power Electronics Systems (CIPS)*, 16–18 March 2010, Nuremberg, German pp. 1–6
23. M. März, P. Nance, Thermal modeling of power electronic systems, in *Proceedings of Application Note Information Technology*, AG Munich, 2000, pp. 1–8
24. H. Agarwal, S. Venugopalan, M.-A. Chalkiadaki, N. Paydavosi, J.P. Duarte, S. Agnihotri et al., Recent enhancements in BSIM6 bulk MOSFET model, in *Proceedings of IEEE International Conference Simulation Semiconductor Processes and Devices*, Glasgow, United Kingdom, September 2013, pp. 53–56
25. S. Russo, V. d'Alessandro, N. Rinaldi, Development of an enhanced ADS electrothermal simulation tool for RF circuits, in *Proceedings of MOS-AK/GSA Workshop*, Rome, Italy, 2010, pp. 1–26
26. R. Hefner, D.L. Blackburn, Thermal component models for electrothermal network simulation. IEEE Trans. Compon. Packag. Manuf. Technol. Part A **17**(3), 413–424 (1994)
27. P.A. Mawby, P.M. Igic, M.S. Towers, New physics-based compact electro-thermal model of power diode dedicated to circuit simulation, in *Proceedings of IEEE Circuits Systems ISCAS*, Sydney, NSW, Australia, Vol. 3, May 2001, pp. 401–404
28. S. Wünsche, C. Clauss, P. Schwarz, F. Winkler, Electro-thermal circuit simulation using simulator coupling. IEEE Trans. Very Large Scale Integr. (VLSI) Syst. **5**(3), 277–282 (1997)
29. V. Košel, R. Illing, M. Glavanovics, A. Šatka, Non-linear thermal modeling of DMOS transistor and validation using electrical measurements and FEM simulations. Microelectron. J. **41**(12), 889–896 (2010)
30. W.V. Petegem, B. Geeraerts, W. Sansen, Electrothermal simulation and design of integrated circuits. IEEE J. Solid-State Circuits **29**(2), 143–146 (1994)
31. V. Košel, S. de Filippis, L. Chen, S. Decker, A. Irace, FEM simulation approach to investigate electro-thermal behavior of power transistors in 3-D. Microelectron. Rel. **53**(3), 356–362 (2013)
32. M. Vellvehi, X. Jorda, P. Godignon, J. Millan, Electro-thermal simulation of a DC/DC converter using a relaxation method, in *Proceedings of 7th International Conference on Thermal, Mechanical Multiphysics Simulation Experiments in Micro-Electronics and Micro-Systems*, Como, Italy, 2006, pp. 1–7
33. R. Gillon, P. Joris, H. Oprins, B. Vandevelde, A. Srinivasan, R. Chandra, Practical chip-centric electro-thermal simulations, in *Proceedings of 14th International Workshop Thermal Investigation of ICs and Systems*, Rome; Italy, 2008, pp. 220–223
34. P. Joris, R. Gillon, A. Srinivasan, R. Chandra (2007), Full-chip electro-thermal simulation using loosely coupled electrical and thermal simulators [Online]. Available: http://www.cadence.com/rl/Resources/conference_papers/9.9Presentation.pdf
35. R. Poore, M. Petersen (2012), Integrated electrothermal solution delivers thermally aware circuit simulation [Online]. Available: https://ieeetv.ieee.org/conference-highlights/ims-2012-microapps-ntegrated-electrothermal-solution-delivers-thermally-aware-circuit-simulation-rick-poore-agilent-eesof
36. S. Filippis, V. Košel, D. Dibra, S. Decker, H. Köck, A. Irace, ANSYS based 3D electro-thermal simulations for the evaluation of power MOSFETs robustness. Microelectron. Reliab. **51**(9), 1954–1958 (2011)
37. A. Chvála, D. Donoval, J. Marek, P. Príbytný, M. Molnár, M. Mikolášek, Fast 3-D electrothermal device/circuit simulation of power superjunction MOSFET based on SDevice and HSPICE interaction. IEEE Trans. Electron Devices **61**(4), 1116–1122 (2014)
38. TCAD Sentaurus User Manual, Version I-2013.12, Synopsys, San Jose, CA, USA, 2013

39. C.-S. Yun, X. Xu, A. Terterian, T. Cilento, Package reliability analysis with coupled electro-thermal and mechanical modeling, in *Proceedings of Materials Research Society Symposium*, San Francisco, CA, United States, vol. 1559, 2013, pp. 54–59
40. F. Nallet, L. Silvestri, C.-S. Yun, S. Holland, M. Rover, T. Cilento, TCAD simulation methodology for electrothermal analysis of discrete devices including package, in *Proceedings of 26th International Symposium Power Semiconductor Devices ICs*, Waikoloa, HI, United States, June 2014, pp. 334–337
41. A. Chvála, D. Donoval, A. Šatka, M. Molnár, J. Marek, P. Príbytný, Advanced methodology for fast 3-D TCAD device/circuit electrothermal simulation and analysis of power HEMTs. IEEE Trans. Electron Devices **62**(3), 828–834 (2015)
42. Synopsys, San Jose, CA, USA (1986), Technology computer-aided design [Online]. Available: http://www.synopsys.com/Tools/TCAD/Pages/default.aspx
43. K. Fischer, K. Shenai, Dynamics of power MOSFET switching under unclamped inductive loading conditions. IEEE Trans. Electron. Devices **43**(6), 1007–1015 (1996)
44. T. Fujihira, Theory of semiconductor super junction devices. Jpn. J. Appl. Phys. **36**(10), 6254–6262 (1997)
45. P. Moens, F. Bogman, H. Ziad, H. De Vleeschouwer, J. Baele, M. Tack et al., UltiMOS: A local charge-balanced trench-based 600V super-junction device, in *Proceedings 23rd International Symposium Power Semiconductor Devices ICs*, 2011, pp. 304–307
46. A. Laprade et al. A New PSPICE Electro-Thermal Subcircuit For Power MOSFETs, Application, Note 7532, Fairchild, 2003
47. P.E. Bagnoli, Thermal resistance analysis by induced transient (TRAIT) method for power electronic devices thermal characterization—Part I: fundamentals and theory. IEEE Trans. Power Electron. **13**(6), 1208–1219 (1998)
48. R. Jancke, A. Wilde, R. Martin S. Reitz, P. Schneider, Simulation of electro-thermal interaction, in *Proceedings Electronics System Integration Technology Conference*, Berlin, Germany, 2010, pp. 1–6
49. Virginia Semiconductor, Inc. Fredericksburg, VA, USA (1997), Basic mechanical and thermal properties of silicon [Online]. Available: http://www.virginiasemi.com/
50. (2013). Thermal conductivity of some common materials and gases [Online]. Available: http://www.engineeringtoolbox.com/thermal-conductivity-d_429.html
51. J. Rhayem, A. Wieers, A. Vrbicky, P. Moens, A. Villamor-Baliarda, J. Roig et al., Novel 3D electrothermal robustness optimization approach of super junction power MOSFETs under unclamped inductive switching, in *Proceedings of 28th IEEE SEMI-THERM Symposium*, San Jose, California, United States, March 2012, pp. 69–74
52. Analog Extensions to Verilog HDL, Version 1.0, Verilog-A Language Reference Manual, Silicon Valley, CA, USA, August 1996
53. A. Chvala, D. Donoval, J. Marek, P. Pribytny, M. Molnar, Power transistor models with temperature dependent parasitic effects for SPICE-like circuit simulation, in *Proceedings International Conference Microelectronics*, Nis, Serbia, 2012, pp. 255–258
54. A. Chvala, D. Donoval, J. Marek, P. Pribytny, M. Molnar, Advanced methodology for fast 3-D TCAD electrothermal simulations of power devices, in *Proceedings of International Conference on Advanced Semiconductor Devices and Microsystems*, Smolenice, Slovakia, 2014, pp. 331–334
55. D.L. Blackburn, Power MOSFET failure revisited, in *Proceedings of 19th Annual IEEE Power Electronics Specialists Conference*, Kyoto, Jpn, April 1988, pp. 681–688
56. Vishay Siliconix. Santa Clara, CA, USA (1994), Unclamped inductive switching rugged MOSFETs for rugged environments [Online]. Available: http://www.vishay.com/docs/70572/70572.pdf
57. NPX Semiconductors, Buck converters for SSL applications, in *Appl. Note*, http://www.nxp.com/documents/application_note/AN10876.pdf
58. A. Chvála, D. Donoval, L. Nagy, J. Marek, P. Príbytný, M. Molnár, 3-D Electrothermal device/circuit simulation of DC–DC converter module in multi-die IC, in *Proceedings European Solid-State Device Research Conference*, Venezia Lido, Italy, 2014, pp. 130–133

# Chapter 6
# MEMS System-Level Modeling and Simulation in Smart Systems

**Gerold Schröpfer, Gunar Lorenz, Arnaud Krust, Benoît Vernay, Stephen Breit, Alexandre Mehdaoui, and Alessandro Sanginario**

## 6.1 Introduction to MEMS System-Level Modeling

### *6.1.1 The Need for System-Level Models for Microsystems*

Most micro-electro-mechanical systems (MEMS) are comprised of a MEMS sensing or actuation element (the "MEMS device"), which is distinct from the accompanying electronics that process the output signal from the device and/or control the device. It's inherent to MEMS that their small scale leads to extremely tight coupling of multi-physics aspects arising from the heterogeneous technologies composing the microsystem. This intimate coupling presents unique challenges for the modeling and design community. The MEMS device itself poses complex design problems that are at best difficult and time consuming to solve with standard field solver analysis. Especially, in the case of transient analysis, many problems are intractable using the currently available finite element or boundary element software.

An additional challenge poses the key to successful system design: the ability to simulate and analyze the dependencies and interaction of its individual components. MEMS product development within any organization divides the development similarly: the MEMS engineers who design the MEMS device, and IC engineers who design the surrounding sensing or control electronics. MEMS engineers and integrated circuit (IC) designers are faced with the need to co-simulate their MEMS and IC designs in a common simulation environment. Co-simulation is required to

---

G. Schröpfer (✉) • G. Lorenz • A. Krust • B. Vernay • S. Breit • A. Mehdaoui
Coventor, Paris, France
e-mail: gschropfer@coventor.com; arnaud.krust@coventor.com; sbreit@coventor.com

A. Sanginario
IIT, Torino, Italy

 
N. Bombieri et al. (eds.), *Smart Systems Integration and Simulation*,
DOI 10.1007/978-3-319-27392-1_6

verify the IC design and to predict yield sensitivity to manufacturing variations. The most obvious path is to do the co-simulation in the environment used by the IC designers, which requires that the MEMS designers deliver a behavioral model of the MEMS device expressed in a hardware description language (HDL) such as Verilog-A or VHDL-AMS. MEMS/IC co-simulations are expected to run with simulation speed comparable to lumped-parameter behavioral models while preserving the accuracy associated with detailed finite-element or boundary-element models. Today, MEMS engineers have only a few choices to deliver behavioral models in these formats. In practice, the employed methods are to handcraft a model, usually in the form of a lookup table, to generate a reduced-order model from finite element analysis (FEA) or to use an existing library of predefined MEMS component models.

### 6.1.2 Compact Model Creation from FEM Using Model-Order Reduction

A common technique for creating fast dynamic models that can be easily inserted into an analog/mixed-signal circuit design environment such as Cadence Virtuoso® or MATLAB Simulink® is called model-order reduction (MOR). An often used MOR technique relies on the principle of modal superposition. In this conventional approach, a first, purely mechanical reduced-order-model (ROM) is directly derived from a detailed mechanical FEM representation. While some research has been published to include mechanical nonlinearities [1, 2], purely linear mechanical ROMs remain common practice. After the mechanical extraction, electrostatic effects are added to the ROM to account for electromechanical coupling. The required electrostatic and fluidic forces and capacitive outputs are commonly represented as polynomial functions. They are extracted from a series of electrostatic FEM analyses on individual comb fingers and electrode segments [3]. While the initial mechanical MOR can be automated to a large extent, tying time-dependent electrostatic loads to modal displacements requires design-dependent engineering judgment and is difficult to generalize or automate. Furthermore, while MOR from mechanical FEM guarantees accurate preservation of the sensor's mechanical Eigen frequencies, the ad hoc approach to including electromechanical coupling relies on various simplifying and error prone assumptions such as multidimensional curve fitting of rational functions and the neglect of nonlinear electromechanical coupling effects such as electrostatic spring softening.

### 6.1.3 Schematic-Driven MEMS Modeling

The alternative approach to MOR from FEM can be summarized as schematic-driven or library-based MEMS modeling. The development of MEMS component

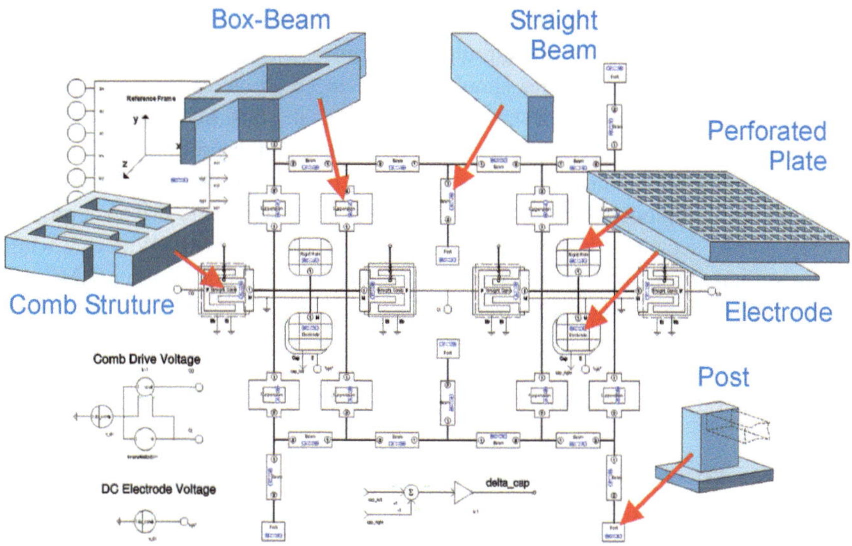


**Fig. 6.1** Schematic of a gyroscope assembled from MEMS components from a behavior model library [7]

libraries for circuit simulators started in the 1990s [4–6] and some of these libraries are now available in commercial tools.

Early on, lumped-element libraries for MEMS closely resembled their electronic counterparts. Like IC designers, MEMS designers needed to work in a schematic-driven environment using symbols that represent individual parametric building blocks or components. These symbols were connected in the schematic to represent a three-dimensional MEMS device, as shown in Fig. 6.1.

The schematic symbols and the mathematical models that they represent enable the exploration of the parametric design space in seconds or minutes with accuracy that rivals conventional FEA [8, 9]. Schematic-driven MEMS design environments, while extremely fast, parametric, and capable of incorporating nonlinear effects, still face obstacles to widespread adoption.

First, creating three-dimensional geometry using symbols and wires is laborious and non-intuitive. MEMS designers, who are responsible for device model creation, traditionally prefer to use either 2-D layout or 3-D mechanical CAD tools for design entry. Therefore, using schematic-driven design entry instead requires a fundamental change in their preferred way of working.

Second, IC designers and system architects, the principal clients for MEMS behavioral models, rely on either signal-flow simulators, such as MATLAB Simulink, or custom IC design and simulation environments from Cadence®, MentorGraphics®, or Synopsys®. None of the standard electronic design automation (EDA) environments provide a particularly attractive environment for MEMS design and, more importantly, there is no standard way to exchange behavioral models between signal-flow and circuit simulators.

Ideally, the MEMS designer should be able to create and modify the MEMS design in a 3-D physical design environment that suits his needs, and then automatically generate required simulation models and layout views for either signal-flow or circuit simulators (hence the appeal of MOR from FEM). On the other hand, simulation models should be parametric (as in the schematic-driven approach) and of course accurately capture the complex behavior of the MEMS device while being sufficiently computationally efficient to allow simulation of the MEMS and IC together with reasonable CPU time.

### 6.1.4 A 3-D Parametric Library for MEMS Design

Coventor's MEMS+® design platform was developed to address the previously described challenges. The MEMS+ platform [10] is a unique blend of various existing technologies including FEA, state-of-the-art behavioral modeling, and model-order reduction. With MEMS+, MEMS designers can work in a 3-D environment that suits their needs and yet easily deliver parameterized behavioral models that are compatible with IC design and system simulation environments. The IC or system designer, meanwhile, will see no difference between including a MEMS device and any other analog or digital component.

The MEMS+ design methodology is based on a library of parametric building blocks that the MEMS designer uses to assemble the desired geometry. Assembled components are automatically connected based on their geometric locations. As an example, Fig. 6.2 shows a three-axis angular rate sensor assembled out of more than 600 individual MEMS+ components [11].

Instead of creating an abstract schematic diagram, the user selects a component from the library, enters values for its parameters, and a corresponding 3-D view is immediately presented on the canvas. This direct creation of a MEMS device in a 3-D view has proven to be much more natural to MEMS engineers compared to the schematic-based approach shown in Fig. 6.1. Furthermore, providing a graphical design entry interface that is separated from the actual system simulation environment allows for alternative design creation methods, including assisted 3-D geometry creation from 2-D layout import, free-hand drawing capabilities and schematic assembly based on Python or MATLAB scripting which can be employed in parallel to a pure library-driven design approach.

The resulting 3-D view differs from traditional 3-D CAD modeling tools in that there are underlying high-order finite elements or sophisticated behavioral models associated with each MEMS building block. The underlying physical representation of each library component can be configured and optimized for a given application. The gyroscope of Fig. 6.2, for example, is internally represented by fully coupled nonlinear system matrices with about 4000 mechanical and 10 electrical degrees-of-freedom (DoF) as well as 3 angular velocity inputs and 9 capacitance outputs.

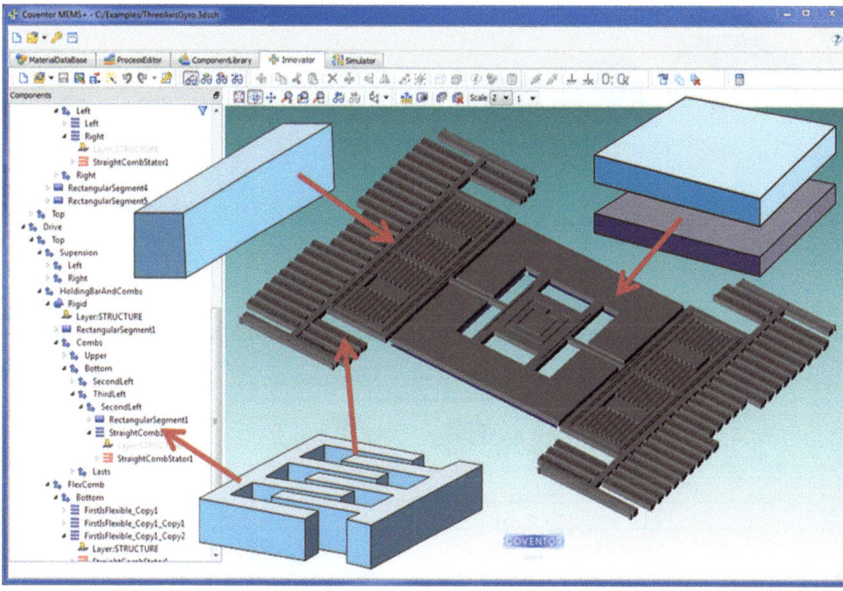


**Fig. 6.2** 3-D schematic of a gyroscope (original design is courtesy of Murata Electronics Oy) assembled with MEMS+® library components including straight beams, rectangular and triangular plates, combs and electrodes [11]

### *6.1.5 MEMS Model Library*

The MEMS+ component library is the product of many years of effort and can be thought of as the MEMS equivalent of the BSIM (Berkeley Short-channel IGFET Model) library in the IC design world. The MEMS+ component library is strictly hierarchical and consists of two distinct groups: basic structural components and component add-ons. Structural components are shapes that can be drawn in two dimensions, with the third dimension being the thickness of one or multiple material layers from which the component is formed. The structural components are basic shapes like rectangles, quadrilaterals, arcs, and pies that can be combined to compose complex flexible structures, as shown in Fig. 6.3.

Flexible mechanical components offer the user one or more underlying model choices. Depending on the given geometry, the user may choose a brick, a shell, a Timoshenko, or a Bernoulli beam element to represent the corresponding structure. Each model choice involves a particular trade-off between simulation accuracy and speed. All mechanical models include support for process-relevant effects, such as sidewall angles and residual stress, multi-layer construction, perforations as well as nonlinear behavior, such as buckling. If technology permits, the user can also assign optional physics models to capture piezo-electric or magnetic effects.

The second group of MEMS+ library components is add-on components which are used to "decorate" the mechanical components with gaps or electrostatic comb

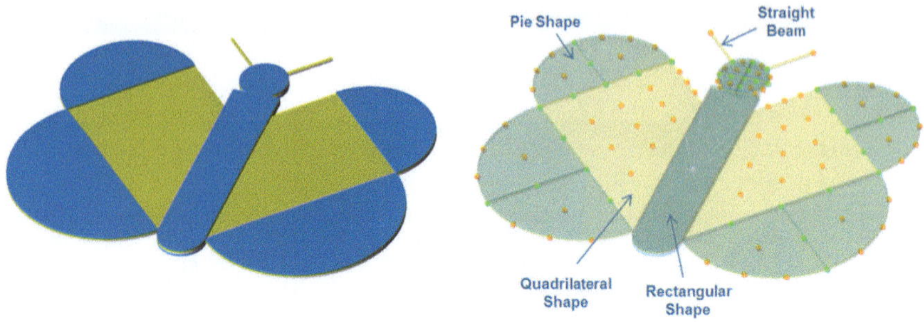


**Fig. 6.3** Complex geometries composed from basic shapes in the MEMS+® component library

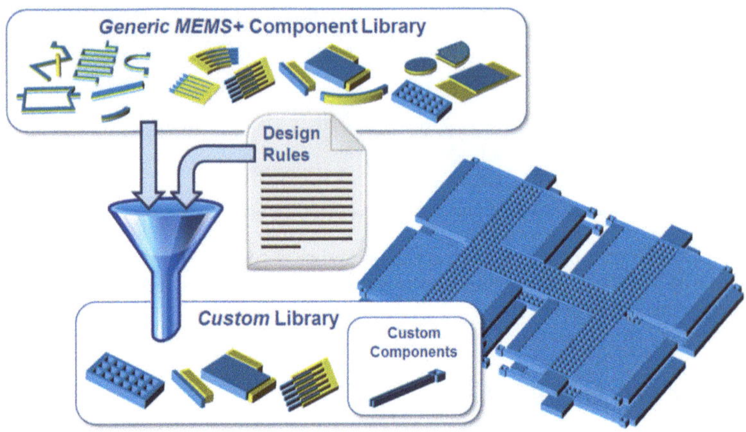


**Fig. 6.4** MEMS+® custom library as part of a MEMS PDK

drives. Similar to the mechanical components, gap components feature optional model choices used to include electrostatic actuation, detection, contact, or squeeze film damping. The corresponding underlying models are based on various different modeling techniques, including analytic formulae, numerical integration, conformal mapping, and finite elements.

### *6.1.6 Process Design Kits for MEMS*

Similar to Process Design Kits (PDKs) known in the IC design world, the MEMS+ component library can be customized for a given MEMS fabrication technology. Library customization is used to eliminate components or modeling choices unfit for a given technology, associate components with certain material layers and impose design rules such as suspension width limits, minimal comb finger spacing, or enforce a certain perforation pattern, see Fig. 6.4.

A customized MEMS+ library might be supplied by an independent MEMS foundry as part of a MEMS PDK [12] or created in-house to impose design rules, eliminate design errors, and encapsulate IP blocks or simply to facilitate component-based design entry.

### *6.1.7 Integration with System Simulators*

All MEMS designs created with MEMS+ can be directly analyzed within the MEMS+ user interface. Similar to standard FEM-based simulation tools, MEMS+ provides simulation and analysis as well as 3-D result visualization capabilities including coupled multi-physics DC (equilibrium), AC (harmonic), modal, and electrostatic pull-in analysis.

The real advantage of using MEMS+, however, is its ability to interface with external system simulation tools. MEMS+ provides two native interfaces which allow the simulation of the full MEMS+ model inside MATLAB Simulink or Cadence Virtuoso see Fig. 6.5. Furthermore, its ability to export geometry in standard 2-D (GDS II) and 3-D (ACIS) formats allows for a seamless link to layout tools or detailed stress and damping analysis with existing FEM/BEM tools.

MATLAB and Simulink from The MathWorks are well known across all engineering disciplines as powerful tools for engineering innovation. These tools allow

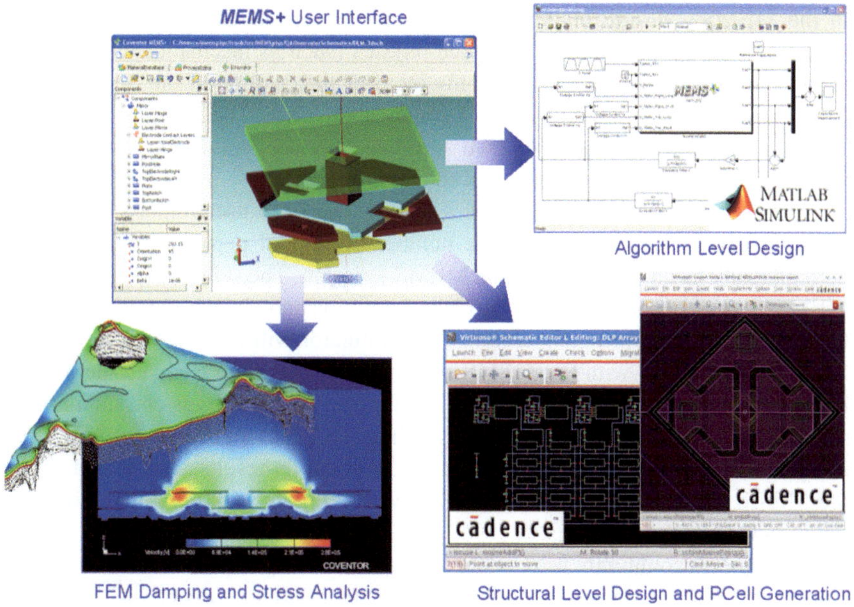


**Fig. 6.5** MEMS+ 3-D design entry and interface options

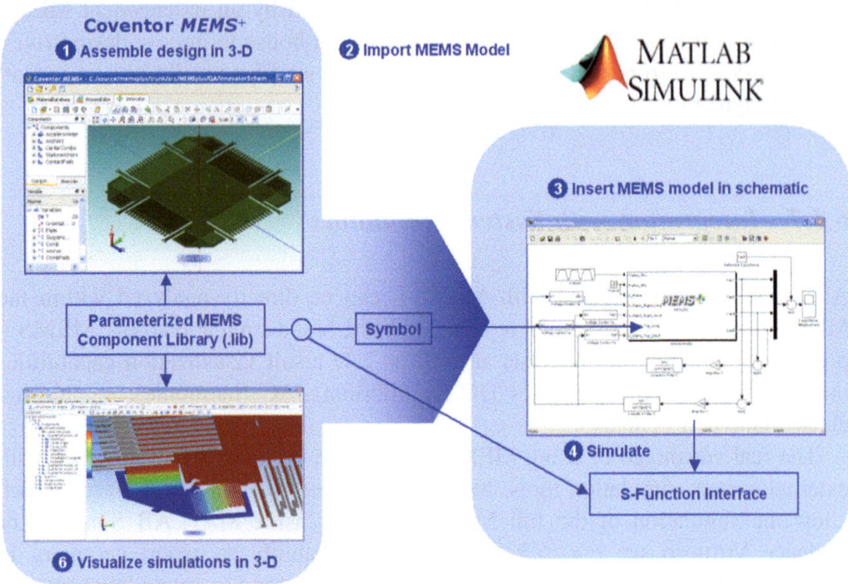


**Fig. 6.6** MEMS+ integration with MATLAB Simulink

engineers to define system models specific to their domain and then simulate their behavior. With MEMS+, designers can import the parameterized model created in the MEMS+ design platform directly into MATLAB or Simulink. MEMS+ requires neither programming of device physics (such as mechanical equations or capacitance extraction) nor FEA; it only requires the creation of the 3-D design in MEMS+ using the tool's intuitive 3-D graphical interface. The MEMS+ MATLAB interface supports both the MATLAB scripting interface and device model import into the Simulink schematic editor, as shown in Fig. 6.6.

MEMS+ automatically generates a symbol, and the user imports it into the Simulink model editor window. The number of symbol ports and parameters is automatically taken from the original MEMS+ design. The symbol representing the MEMS+ model can be inserted into larger systems and simulated with the native MATLAB solvers. During simulations, the simulator will connect, via Simulink's S-function interface, with the MEMS+ component library to evaluate the MEMS behavioral model at each time step. In addition to standard transient simulations, MEMS+ provides additional analyses such as DC, DC transfer, modal, and AC analysis. On completion of a simulation, the simulation results can be loaded into the MEMS+ user interface and visualized via *X*–*Y* graphs and as fully contoured three-dimensional animations.

IC engineers commonly use Cadence Virtuoso to design the analog/mixed-signal electronics that accompany a MEMS device. In order to succeed, IC engineers require fast and accurate models of the MEMS device in the Cadence model library. MEMS+ facilitates the required model exchange by providing an easy way to

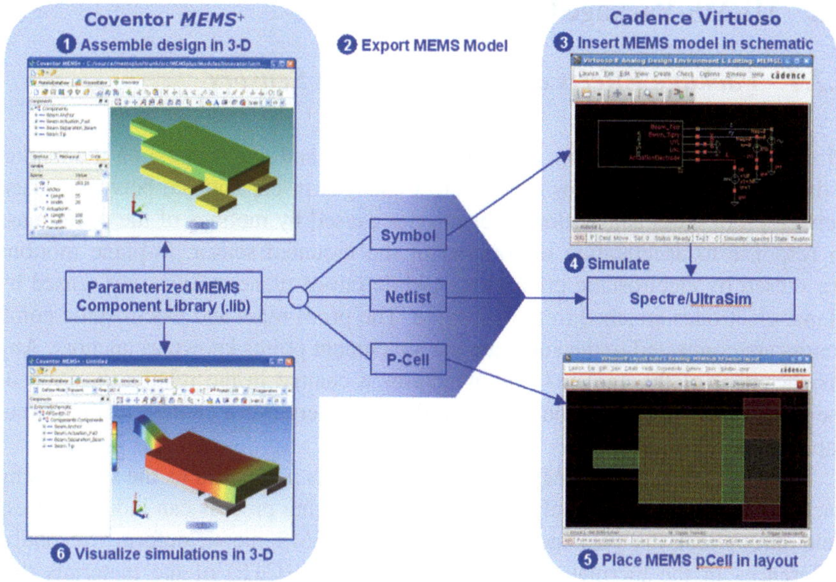


**Fig. 6.7** MEMS+ 3-D design entry and interface options

import MEMS+ models to the Cadence model library; see Fig. 6.7. Every device created in MEMS+ can be imported into the IC design environment in the form of a netlist and a schematic symbol. Similar to the MEMS+ MATLAB Simulink interface, the number and names of the pins on the schematic symbol are controlled by the MEMS engineer and represent electrical connections to the MEMS device. The MEMS symbol can be placed in a schematic in the IC schematic editor and surrounded by the complete IC design.

Simulations can be run in any of the Cadence circuit simulators that are compatible with Virtuoso, including Spectre, SpectreRF, and APS. The simulator will connect with the MEMS component library to evaluate the MEMS model at each simulation point, i.e., time step or frequency. It is important to highlight that all external solvers supported by MEMS+ (including MATLAB) use the same component library during the actual simulation. All MEMS+ supported simulations are therefore expected to be of comparable accuracy.

On completion of a simulation, the designer can view the simulation results in the MEMS+ 3-D viewer, which can animate the motion of the MEMS device. At any time, but especially when the MEMS and IC designers are satisfied with the MEMS design, they can export a parameterized layout cell (PCell) that can generate a layout of the MEMS device.

## 6.2 MEMS-Package Co-Design

### *6.2.1 Package Effects on MEMS Motion Sensors*

Most MEMS motion sensors (both accelerometers and gyroscopes) on the market today are capacitive devices. They detect changes in capacitance between a suspended proof mass and stationary electrodes caused by motions of the proof mass in response to acceleration or rotation of the mounted sensor. In-plane motions are sensed by electrostatic comb drives while out-of-plane motions are sensed by planar electrodes attached to the substrate. The proof mass and electrostatic comb stators are connected to the substrate via attachment points known as anchors. Any displacement of the anchor locations causes a change in the relative position of the proof mass to the stationary electrodes and therefore a change in the output capacitance of the sensor.

The output of capacitive MEMS motion sensors is influenced by the surrounding package. Firstly, thermal cycling during the packaging process can induce a stress field throughout the package that causes the MEMS die to deform. Secondly, differences in the thermal coefficient of expansion (CTE) of the materials that comprise the package and die can cause the package and die to deform as the ambient temperature changes. We refer to the underlying silicon die, on which the MEMS device is fabricated, as the substrate. Substrate deformation can change the position and orientation of the anchors enough to affect the shape and relative distance between the MEMS parts. For instance, substrate deformation can alter the separation between a moving layer and an electrode attached to a substrate, thus changing the capacitance between them. Critical MEMS performance specifications, such as zero offset in accelerometers and drift bias in gyros, may therefore vary with substrate deformation and, in turn, ambient temperature.

### *6.2.2 Methodology for MEMS-Package Co-Simulation*

The methodology involves three steps. The first step is to simulate the static deformation of the package and MEMS substrate for a range of temperature values. This simulation is usually performed by package designers using a general-purpose FEA tool. Note that the package analysis includes the MEMS substrate and may even include the PC board to which the package is attached, but not the MEMS device itself. The MEMS device is very small compared to the substrate and the package and therefore assumed to have negligible effects on the deformation of the latter. This is an important simplifying assumption because the MEMS device has a lot of structural details that would greatly increase the computing resources required for the FEA. Second, the package simulation results are imported into Coventor's MEMS+ design environment where they are mated with the MEMS device. MEMS+ moves the anchor and electrode locations in accordance with the package results, using linear interpolation between temperature values in the

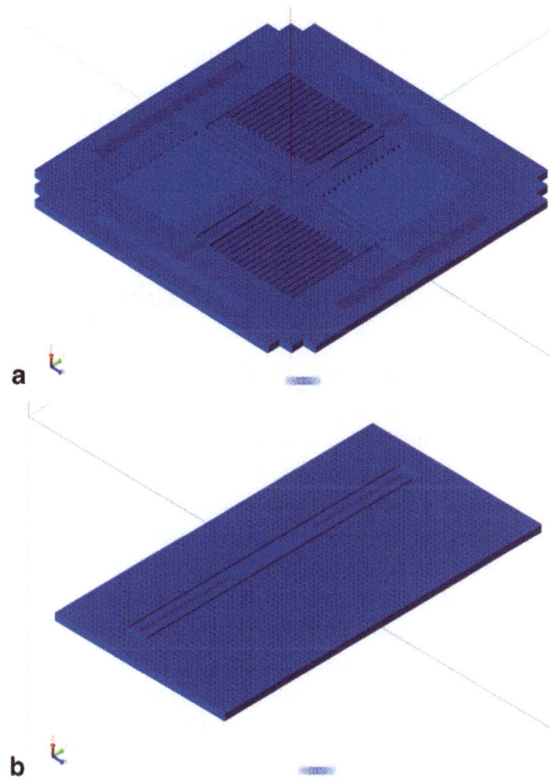

**Fig. 6.8** MEMS+ 3-D schematic models of the two sensing elements composing a 3-axis accelerometer, one element (**a**) sensitive to in-plane motion (*X*, *Y*) and another (**b**) to detect out-of-plane motion (*Z*). Original design is courtesy of STMicroelectronics Srl

FEA results. Third, the MEMS+ model of the device with displaced locations is simulated in MATLAB, Simulink, or Cadence Virtuoso to predict the temperature stability of the sensor output.

### *6.2.3 3-Axis MEMS Accelerometer Example*

We apply the methodology to a 3-axis capacitive accelerometer [13] consisting of two sensing elements: one sensitive to in-plane motion (*X*, *Y* directions) and the other sensitive to out-of-plane motion (*Z* direction). Figure 6.8 shows a 3-D model of the accelerometer in MEMS+, created by assembling individual MEMS+ library components including beams, perforated plates, comb fingers, and electrodes.

The principle of operation for the *X*–*Y* sensing part is as follows: the accelerometer can be divided into four quadrants. Each section has a sensing element composed of comb electrodes attached to the movable perforated mass and their respective stator electrodes that are fixed. When the structure supporting the device is subjected to acceleration in the *X*- or *Y*-directions, the movable mass electrodes on one side move closer to their stators while those on the other side move further away. The differential capacitance between the two sides is linearly proportional to the acceleration of the supporting structure. Similarly, the $Z$ acceleration of the supporting structure is translated into a change of capacitance. When a positive $Z$ acceleration is applied the proof mass will move toward the underlying electrodes increasing the capacitance. In the case of negative $Z$ acceleration the mass will move away from the electrodes thus decreasing the capacitance.

### *6.2.4 Results of MEMS-Package Co-Simulation*

Following the first step of the methodology, the package model is created and deformed through a parametric thermo-mechanical analysis using FEA, in this case with CoventorWare®. The temperature values for the parametric analysis range from −100 °C to 100 °C in multiple steps. The meshed model of the package prepared for thermo-mechanical simulation is shown in Fig. 6.9a. In the second step, the simulation results of the deformed package are imported into MEMS+ to be linked with the accelerometer model. Figure 6.9b represents the accelerometer (blue colored) placed in its package (green colored). Third, we apply acceleration under different temperature values to evaluate the package influence not only in the static case but also in the dynamic one. Figure 6.9c shows the perforated mass displacement for temperatures ranging from −100 °C to 100 °C with a 10 °C step. We can see the influence of the package on the displacement of the $XY$ accelerometer.

The methodology allows the prediction of package effects on critical sensor outputs such as sensitivity or zero-offset. Figure 6.10 shows the zero g offset with temperature for the $Z$-axis accelerometer. Furthermore, the methodology is fully compatible with the reduced-order modeling approach described in the next section. This means the MEMS model with package effects can be run in it is native format using the full MEMS+ system matrix or with the reduced number of DoF of a reduced-order model which, consequently, significantly reduces the simulation time, especially for transient simulations.

## 6.3 Reduced-Order-Modeling and Verilog-A Extraction for MEMS

### *6.3.1 Introduction*

While MEMS/IC co-simulation is possible with the native MEMS+® third-party solver interfaces, transient simulations of large models as shown in Fig. 6.2 are likely to be too slow for routine MEMS/IC co-design. Although the MEMS+

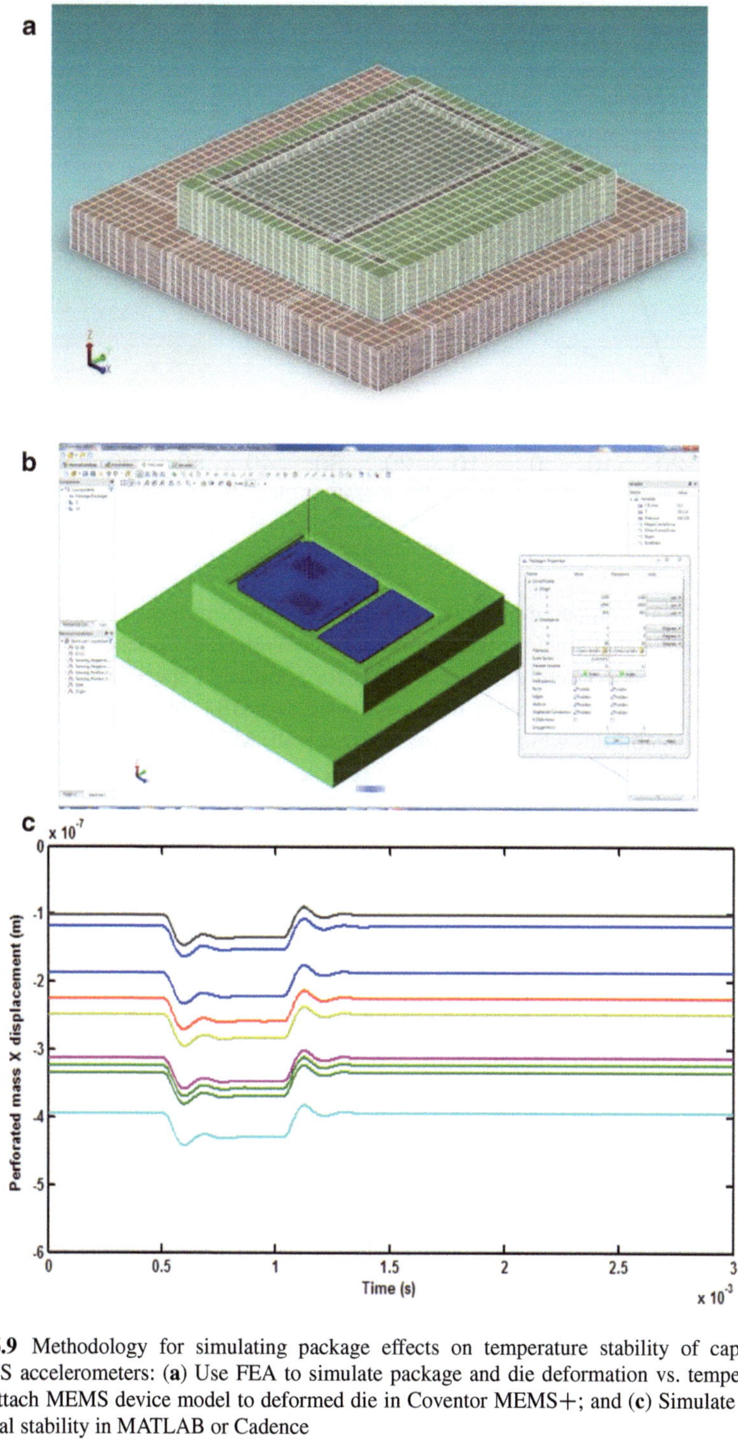


**Fig. 6.9** Methodology for simulating package effects on temperature stability of capacitive MEMS accelerometers: (**a**) Use FEA to simulate package and die deformation vs. temperature; (**b**) Attach MEMS device model to deformed die in Coventor MEMS+; and (**c**) Simulate sensor thermal stability in MATLAB or Cadence

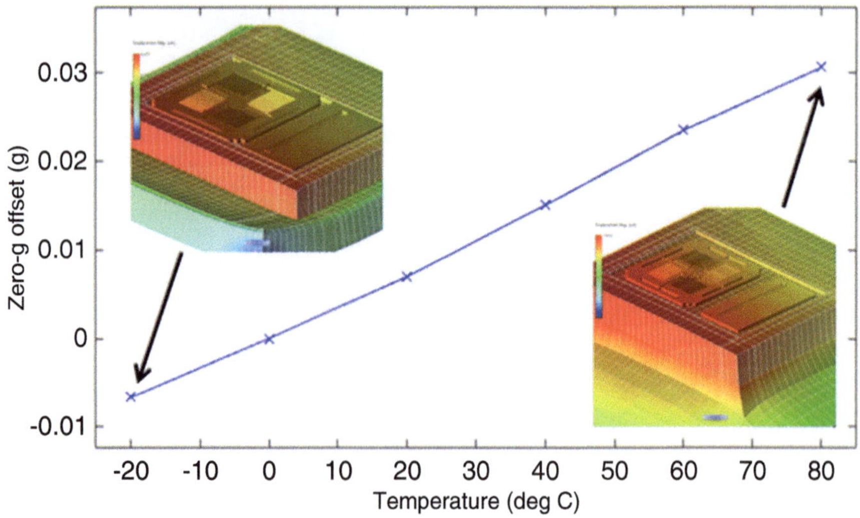


**Fig. 6.10** Simulated zero g offset with temperature

models are tiny compared to a conventional FEM, they may still be large enough to present a big challenge for the supported MathWorks and Cadence solvers.

To address the system designer's needs for simulation speed while preserving the critical nonlinear characteristics of a device, MEMS+ offers additional model extraction capabilities which allow the original (full) MEMS+ model to be reduced to a lower number of degrees-of-freedom (DoF).

### *6.3.2 Reference Model Creation*

As described above, the MEMS+ approach yields a compact, nonlinear, multi-physics model consisting of interconnected rigid plates, beam elements, and shell elements as well as electromechanical transducers representing electrodes and comb fingers [10]. MEMS+ automatically assembles the individual component models into a *single* multi-physics system of equations [14]:

$$F\left(X,\dot{X},\ U\right) = 0, \tag{6.1}$$

where $X \in \mathbb{R}^n$ is the state vector with all mechanical and electrical DoF. The input vector $U \in \mathbb{R}^m$ includes all input voltages, external accelerations, and angular velocities; $F : \mathbb{R}^n \times \mathbb{R}^n \times \mathbb{R}^m \to \mathbb{R}^n$ represents the sum of forces.

### 6.3.3 Model-Order Reduction

The linearized version of the reference system (6.1) around a specific operating point $X_0, \dot{X}_0, U_0$ is commonly written as

$$E\dot{X} = AX + BU + R, \tag{6.2}$$

with $A = \partial_X F \in \mathbb{R}^{n \times n}$, $E = \partial_{\dot{X}} F \in \mathbb{R}^{n \times n}, B = \partial_U F \in \mathbb{R}^{n \times m}$ and $R = F\left(X_0, \dot{X}_0, U_0\right) - AX_0 - BU_0 + E\dot{X}_0$. The point $X_0, \dot{X}_0$ is computed by solving the reference system with user-specified inputs $U_0$ (e.g., bias voltages or displacement offsets).

Let $V \in \mathbb{R}^{n \times q}$ be the reduction matrix, which supposes that $q < n$ and let $\tilde{X} \in \mathbb{R}^q$ be the vector of reduced states. For an example gyroscope [11, 15], the original $n \approx 4000$ DoF can be reduced to $q \approx 20$ DoF by building an appropriate matrix $V$. Note, the reduced model is only accurate for a given simulation if $\tilde{X}$ satisfies $X \approx V\tilde{X}$.

After replacing $X$ by $V\tilde{X}$ in (6.2), Galerkin projection (left-multiplication by $W \in \mathbb{R}^{q \times n}$) is used to recover a solvable system. For stability reasons, we define $W = V^T$. The resulting linear reduced system is

$$V^T E V \dot{\tilde{X}} = V^T A V \tilde{X} + V^T B U + V^T R,$$

which can be rewritten as

$$\tilde{E}\dot{\tilde{X}} = \tilde{A}\tilde{X} + \tilde{B}U + \tilde{R},$$

with $\tilde{E} = V^T E V$, $\tilde{A} = V^T A V$, $\tilde{B} = V^T B$ and $\tilde{R} = V^T R$.

### 6.3.4 Reduction Matrix

The accuracy and stability of the reduced system greatly depends on the choice of the matrix $V$. The modal superposition method [1] ensures that certain eigenvalues of the system are preserved and stability of the reduced linear model is ensured, as long as the full model is stable. A vector $X$ can be decomposed into

$$X = \sum_i^n \Phi_i x_i,$$

where the $\Phi_i \in \mathbb{C}^n$ are the eigenvectors of the system. Suppose, for simplicity, that we are interested only in the first two modes. Each mode has a real part ($\Re$) and an imaginary part ($\Im$). The columns of the matrix $V$ are then $\Re(\Phi_1)$, $\Im(\Phi_1)$, $\Re(\Phi_2)$, $\Im(\Phi_2)$. For accuracy reasons, we also want to make sure that the linearization point $X_0$ belongs to the space spanned by the columns of $V$ by simply including $X_0$ as a new column of $V$. Doing so, we can no longer guarantee stability of the system. It is however possible to preserve the block structure of the mechanical part of the system with a simple modification of $V$. The full matrices $A$ and $E$ have the structure

$$A = \begin{pmatrix} K & D & * \\ 0 & I & * \\ * & * & * \end{pmatrix} \quad \text{and} \quad E = \begin{pmatrix} 0 & M & * \\ I & 0 & * \\ * & * & * \end{pmatrix},$$

with $K$, $M$, and $D$ the stiffness, mass, and damping matrices. The stars (*) represent the electrical part of the system.

The corresponding state vector can be written as

$$X = \begin{pmatrix} X^p \\ X^v \\ X^e \end{pmatrix},$$

where $X^p$ contains the displacement DoFs, $X^v$ the velocity DoFs, and $X^e$ the electrical DoFs. As the dimension of $X^e$ is very small (only a few DoFs), we only need to reduce $X^p$ and $X^v$. In the same fashion, we define $V^p$ as the first rows of the original matrix $V$. We ensure that ${V^p}^T V^p = I$ with a singular value decomposition (SVD) on $V^p$. Choosing our reduction matrix as

$$V = \begin{pmatrix} V^p & 0 & 0 \\ 0 & V^p & 0 \\ 0 & 0 & I \end{pmatrix},$$

preserves the original block structure of $A$ and $E$:

$$\tilde{A} = \begin{pmatrix} \tilde{K} & \tilde{D} & * \\ 0 & I & * \\ * & * & * \end{pmatrix} \quad \text{and} \quad \tilde{E} = \begin{pmatrix} 0 & \tilde{M} & * \\ I & 0 & * \\ * & * & * \end{pmatrix},$$

with $\tilde{K} = {V^p}^T K V^p$, $\tilde{D} = {V^p}^T D V^p$ and $\tilde{M} = {V^p}^T M V^p$. The stars (*) are reduced as well.

Compared to other methods, the chosen reduction matrix is expensive as it has twice as many columns, leading to twice as many unknowns in the ROM. However, the gain in stability proved to justify the extra computational cost.

### 6.3.5 Higher Order Terms

A wide range of methods can be applied to reduce a full model to a linear model and this field of research is now quite mature. However, a certain number of physical behaviors are nonlinear and cannot be recovered by such a simple model. For example, the electrostatic force is known to be quadratic with respect to voltage. Other examples are the rotating inertial forces (e.g., the Coriolis force), which a linear model cannot represent. Accurately modeling these nonlinearities in a ROM is a challenging task.

Interpolation between different linear models can be used to recover nonlinearities that are added to the ROM, especially the electrostatic softening effect. This approach, although very accurate, tends to create very large models especially if multiple electrodes are involved [15]. A well-known application of interpolation is the trajectory piece-wise linear method [16], which is quite successful at recovering mechanical nonlinearities of the MEMS structure.

Another approach is to add polynomial terms in an analytical or a semi-analytical manner.

Both approaches can successfully be mixed into a single ROM [11].

### 6.3.6 Electrostatic Force

The input vector $U$ can be decomposed into the voltage inputs $U_e$ and all other inputs $U_*$. For a gyroscope, $U_* \in \mathbb{R}^3$ notably contains the angular rates. Decomposing the matrix $\tilde{B}$ accordingly yields $\tilde{B}U = \tilde{B}_e U_e + \tilde{B}_* U_*$.

Electrostatic forces are known to be quadratic terms with respect to voltage and can be written as

$$F_e(X, U_e) = \begin{pmatrix} U_e^T C_1(X) U_e \\ \vdots \\ U_e^T C_n(X) U_e \end{pmatrix},$$

where $C_1, \ldots, C_n$ are the corresponding capacitance matrices. The dependence of these matrices with respect to displacement is nontrivial and needs to be approximated in a ROM. In a first approximation, these matrices can be considered to be constant. This assumption removes cross terms between displacement and voltage which are known to cause the electrostatic softening effect. In order to accurately recover electrostatic softening terms at a lower cost (i.e., without interpolation), we consider the matrices $C_i$ to depend linearly on displacement. They are computed by finite differentiation of $A$ and $B$. With $\tilde{C}_i\left(\tilde{X}\right) = C_i\left(V\tilde{X}\right)$, the electrostatic force term is reduced to

$$\tilde{F}_e\left(\tilde{X}, U_e\right) = V^T \begin{pmatrix} U_e^T \tilde{C}_1\left(\tilde{X}\right) U_e \\ \vdots \\ U_e^T \tilde{C}_n\left(\tilde{X}\right) U_e \end{pmatrix}.$$

The model should no longer contain terms linear with respect to voltage. We also need to remove these terms from the matrix $\tilde{B}$, which corresponds to removing $\tilde{B}_e U_e$.

The resulting system of the reduced model is now

$$\tilde{E}\dot{\tilde{X}} = \tilde{A}\tilde{X} + \tilde{B}_* U_* + \tilde{F}_e\left(\tilde{X}, U_e\right) + \tilde{R}. \tag{6.3}$$

This is enough to ensure that the first derivative with respect to displacement is exact at the point of extraction, yielding a correct softening.

### 6.3.7 Inertial Forces

There are three inertial rotating fictitious forces to consider:

- Centrifugal force $F_{\text{Centrifugal}} = -m\Omega \times (\Omega \times r)$,
- Coriolis force $F_{\text{Coriolis}} = -2m\Omega \times \dot{r}$,
- Euler force $F_{\text{Euler}} = -m\frac{d\Omega}{dt} \times r$,

where $r \in \mathbb{R}^3$ is the distance to the center of rotation and $\Omega \in \mathbb{R}^3$ the angular velocity of the reference frame. Let $F_{\text{inertial}} = F_{\text{Centrifugal}} + F_{\text{Coriolis}} + F_{\text{Euler}}$ and note that $r$ and $\dot{r}$ can be expressed in terms of $X$. This inertial term can be analytically preserved and reduced to

$$\tilde{F}_{\text{inertial}}\left(\tilde{X}\right) = V^T F_{\text{inertial}}\left(V\tilde{X}\right).$$

### 6.3.8 Verilog-A Specific Implementation

The ROM that we have developed in the previous paragraphs is a system of equations that must be converted to satisfy Kirchoff's law.

Each reduced DoF $r_i$ in $\tilde{X}$ is associated with an electrical signal `r_i` declared by `electrical r_i;`. Its voltage `V(r_i)` being the unknown value of $r_i$, while its current `I(r_i)` represents the force applied.

It is sometimes useful to be able to probe a non-reduced DoF, for example, to know how a specific point of the device has moved during the simulation. Conversely, it is sometimes necessary to apply a force or set the position of a point.

Suppose we are interested in the non-reduced DoF $x_i$ in $X$. Let $V_i$ be the $i$th row of $V$. An approximation of $x_i$ can be reconstructed out of the reduced DoFs by $x_i \approx V_i\tilde{X}$.

This is represented in Verilog-A code:

```
I(x_i) <+V(x_i)-(V_i_0 * V(r_0) + V_i_1 * V(r_1)...);
```

Applying a force $f_i$ to the point $x_i$ creates a column $b_i$ in the matrix $B^*$ that is multiplied by $f_i$. For the ROM, this column is reduced to $\tilde{b}_i = V^T b_i$. The corresponding term is added to each equation:

```
I(r_j) <+b_j_i * I(x_i);
```

Thanks to these operations, a pin representing $x_i$ can be created on the Verilog-A module. It allows probing the value (voltage) of the reconstructed DoF when applying a force (current), or vice-versa. In particular, these additional pins can be used to connect external models to the ROM.

## 6.4 Toward Higher Level MEMS System Design

The tight integration of MEMS devices with surrounding electronic and control systems introduces new needs and opportunities in complex system simulation. Co-development of software and hardware subparts may considerably improve the hardware and highly impacts the design of embedded systems [17]. To obtain these improvements, the virtual prototyping methodology must be able to simulate the overall assembly in order to quickly identify technical issues and incompatibilities in early design phases. The following example is based on the SystemC standard and its Analog Mixed-Signal (AMS) extensions. It demonstrates the ability to simulate the behavioral model of MEMS devices, for instance an accelerometer, connected to digital components responsible for the signal processing and the digital conversion to a CPU.

### *6.4.1 HDL-Based Simulation Environment*

HDLs address the design and the modeling of digital integrated circuits (ICs) and aim at defining logic synthesis systems. VHDL and Verilog are the de facto standards in industry, targeting the low-level simulation of ICs at the register transfer level (RTL) or even at the gate level. Due to the increasing number of analog components in ICs, these traditional HDLs were extended to permit both digital and AMS simulations, for instance with VHDL-AMS and Verilog-AMS [18]. Nevertheless, these solutions remain decoupled from the development of software

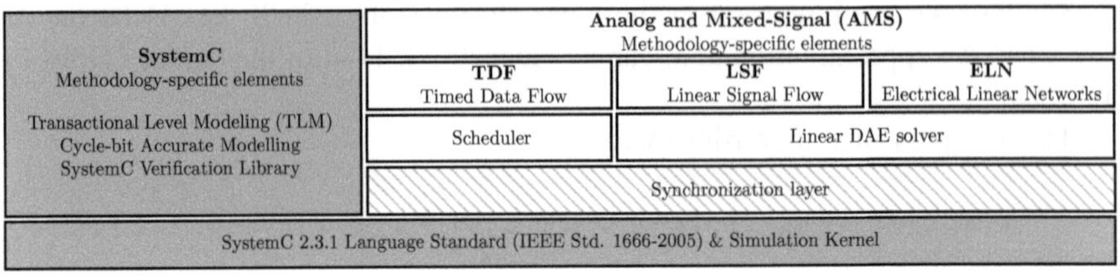


**Fig. 6.11** SystemC-AMS architecture and models of computation definition [21]

running on the targeted architectures. To better address the concurrent development of application-specific software and hardware, higher-level simulation languages, mostly based on C/C++, emerged over the last decade [19].

SystemC [20] is a C++ library for system-level simulation of digital hardware and software. This standard describes a system as an interconnected set of processing entities scheduled by an event-driven simulation kernel. This environment also enables early functional and architectural verification through a dedicated framework such as the Universal Verification Methodology (UVM). Furthermore, AMS extensions have been developed and standardized in SystemC-AMS [21]. This standard defines the synchronization between the discrete event and continuous-time solvers that are responsible, respectively, for digital and analog component simulation. This mechanism enables the definition of additional models of computation (MoCs) [22], each dedicated to a specific physical domain (Fig. 6.11). Nevertheless, SystemC-AMS still has limited efficiency for capturing continuous-time behavior, especially nonlinear behavior [23]. Despite these drawbacks, SystemC-AMS is worthwhile for simultaneously simulating software and hardware with additional analog peripherals, such as MEMS devices. Another potential advantage of SystemC-AMS is its coupling with a verification process like UVM [24].

## 6.4.2 *System-Level Integration*

Integration of a MEMS behavioral model in SystemC-AMS has been achieved with a linearized reduced-order model generated from *MEMS+*®. This model is represented by a state-space system in descriptor form as described in the previous section. Behavioral nonlinearity is reconstructed during simulation, since it depends on the current values of the states and inputs of the system.

The SystemC-AMS standard defines specific modules for the state-space representation (*sca_ss*) in two different MoCs (*sca_tdf*, *sca_lsf*), but they lack sufficient support for matrix operations and nonlinear solving methods. On the one hand, the TDF MoC is used for procedural behavior, i.e., processing samples tagged over the time. The activation schedule of a set of interconnected TDF modules can be statically determined since the number of read and write operations is fixed and known in advance. This MoC is thus well suited for multi-rate signals. On the other

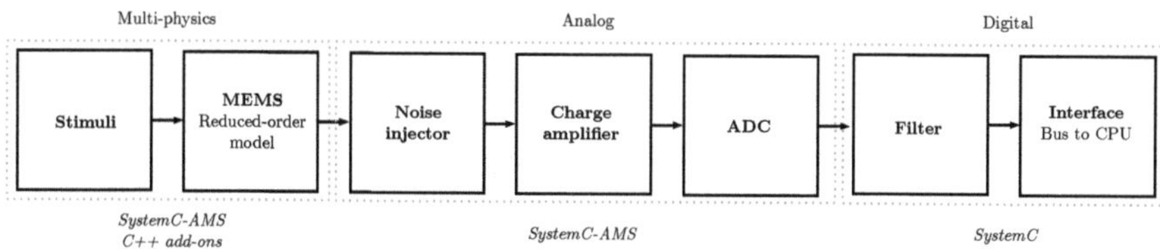


**Fig. 6.12** Integration of a MEMS reduced-order model in a SystemC-based test bench

hand, the LSF MoC is focused on the description of linear, time-invariant, non-conservative systems using block diagram primitives. Switching and modulation of module parameters through external signals are allowed. This makes the LSF MoC more suitable for simple continuous-time controllers and filters. In this study the TDF MoC is preferred since it directly handles the synchronization with SystemC and should benefit from recent improvements [25].

The following example shows the integration of a MEMS behavioral model in a SystemC-AMS test bench. The approach takes advantage of the hardware/software co-simulation implemented in SystemC and the coupling with analog behaviors allowed in SystemC-AMS. The method thus integrates the reduced-order model introduced in the preceding sections and the associated solving methods directly in a SystemC-AMS simulation environment. The reduced-order model is fully accessible through a C++ Application Programmable Interface (API) which allows configuring the model and accessing it during simulation. The inter-connection of the model in the test bench satisfies the SystemC-AMS standard.

### *6.4.3 Test Bench Implementation*

In this example, a MEMS model encapsulated in a TDF module is connected to analog and digital components that implement a traditional digital signal processing (DSP) workflow (Fig. 6.12). The inputs of the microsystem are both mechanical and electrical stimuli, set to analyze the system response to an acceleration step. The MEMS output capacitance is amplified, converted, and filtered through DSP modules finally sent to a CPU through specific bus protocols (I2C, SPI . . . ).

This example demonstrates a valuable simulation environment since the MEMS model can be shared between designers and engineers responsible for its integration. While preserving the integrity and accuracy of the model, this framework allows system engineers to test and verify that the hardware/software application is consistent with regards to external devices, such as sensors.

The test bench has been simulated with an accelerometer similar to the one introduced in the section above. This latter is connected to the corresponding stimuli sources (voltage, acceleration) in order to verify its mechanical response to an acceleration step of amplitude 1g. The same accelerometer behavioral model

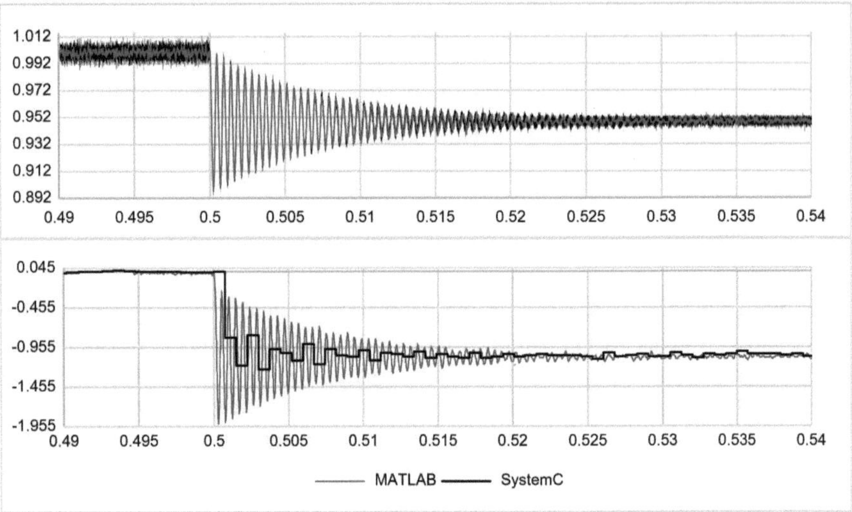


**Fig. 6.13** Simulated accelerometer capacitance variation in response to a *z*-axis acceleration impulse (1g)|Noisy output of the charge amplifier (*top*) and 8-bit filtered signal (*bottom*)

has also been simulated with MATLAB Simulink to verify the accuracy and the performance of SystemC-AMS simulation.

The study first compares the reduced-order model against the full one initially created in MEMS+® in order to verify the accuracy of the method. The relative error of the simulation is fairly low (<1 %) and mainly induced by the ROM process itself, as verified on different simulators.

The integration scheme produces white noise on the output signal, i.e., the capacitance variation. This signal, once amplified and filtered, is converted into an 8-bit signal better suited to further digital processing (Fig. 6.13). This example demonstrates the ability of SystemC-AMS to handle signal processing and bit conversion. Moreover the simulation runs quite fast compared to alternate approaches.

## 6.4.4 Conclusion and Perspectives

The system-level simulation of MEMS devices through an extended version of HDL, such as SystemC-AMS, introduces new capabilities to better address the virtual prototyping of complex systems.

First, the evolution of HDLs tends to handle at early stage the co-development of specific hardware/software applications. This is especially enabled by combining system programming languages, like C/C++, with traditional HDL synthesis methodologies. Furthermore APIs augment the capabilities of simulation software tools by connecting them with other domain-specific environments. Finally modeling methods, like MOR, define lightweight solutions while preserving the accuracy of behavioral models and thus speed up the simulation.

Nevertheless, despite their performance and coherency, monolithic architectures like SystemC-AMS remain extremely limited in terms of extensibility, i.e., the definition of new MoCs, and scalability, since the simulation relies on a single-thread kernel. These drawbacks also highlight the challenges of handling, in a common simulation environment, models describing different physical domains.

Alternatively, co-simulation frameworks combine multiple domain-specific tools and orchestrate the overhaul simulation. Accuracy is preserved, but the performance of such solutions is unacceptable at early design phases. Moreover the interfaces defined by the FMI [26] or HLA [27] standards suffer from a lack of adoption, especially due to their intrusive definition of synchronization and orchestration mechanisms directly inside the simulators.

Focused either on combining models (monolithic architectures) or simulators (co-simulation frameworks), the current solutions introduce novel design and simulation methodologies for integrating complex assemblies at the system level. They mostly rely on the collaboration between expert teams, as shown here for MEMS devices. These methods also call for greater convergence of virtual and physical prototypes through dedicated software solutions that will be even more flexible and connected to better address the complexity of such systems.

## References

1. J. Mehner, Modal-superposition-based nonlinear model order reduction for MEMS gyros, in *System-Level Modeling of MEMS* (Wiley-VCH Verlag GmbH & Co. KGaA, 2013), pp. 291–309
2. M. Dorwarth, S. Kehrberg, R. Maul, R. Eid, F. Lang, B. Schmidt, J. Mehner, Nonlinear model order reduction for high Q MEMS gyroscopes, in *11th International Multi-Conference on Systems, Signals and Devices* (2014)
3. T. Mähne, K. Kehr, A. Franke, J. Hauer, B. Schmidt, Creating virtual prototypes of complex micro-electro-mechanical transducers using reduced-order modelling and VHDL-AMS, in *Proceedings of the 8th International Forum on Specification and Design Languages* (2005)
4. D. Teegarden, G. Lorenz, R. Neul, How to model and simulate microgyroscope systems. IEEE Spectrum **35**, 66–75 (1998)
5. N. Zhou, C.V. Jason, K. Pister, Nodal analysis for MEMS design using SUGAR v0.5, in *MSM 98, Technical Proceedings of the 1998 International Conference on Modeling and Simulation of Microsystems* (Santa Clara, 1998)
6. M.S. Kranz, J.E. Vandemeer, G.K. Fedder, Hierarchical representation and simulation of micromachined inertial sensors, in *MSM98* (Santa Clara, 1998)
7. G. Lorenz, J. Repke, in *Sensors in Automotive Technology*, vol. 4 (Jossey-Bass Publishers (A Wiley Company), 2006), pp. 58–72. ISBN: 978-3-527-60507-1
8. S. Breit, C. Welham, S. Rouvillois, M. Kraft, M. McNie, Simulation environment for accurate noise and robustness analysis of MEMS under mixed-signal control, in *Proceedings of the ASME International Mechanical Engineering Congress* (Boston, 2008)
9. S. Maity, S. Liu, S. Rouvillois, G. Lorenz, M. Kamon, Rapidly analyzing parametric resonance and manufacturing yield of MEMS 2D scanning mirrors using hybrid finite-element/behavioral modeling, in *SPIE Photonics West, MOEMS MEMS Conference* (2014)

10. G. Lorenz, G. Schröpfer, 3D parametric-library-based MEMS/IC design, in *System-Level Modeling of MEMS* (Wiley-VCH Verlag GmbH & Co.KGaA, 2013), pp. 407–424
11. A. Parent, A. Krust, G. Lorenz, I. Favorskiy, T. Piirainen, Efficient nonlinear Simulink models of MEMS gyroscopes generated with a novel model order reduction method, in *Transducers* (2015)
12. J. Doblaski, PDK-based design automation enablement for a surface-micromachined capacitive MEMS foundry process, in *SmartSystems Integration* (Copenhagen, Denmark, 2015)
13. A. Sanginario, A. Mehdaoui, S. Zerbini, G. Schröpfer, New design methodology for MEMS-electronic-package co-design and validation for inertial sensor systems, in *IEEE DTIP 2015 (Design, Test, Integration and Packaging of MEMS/MOEMS)* (Montpellier, France, 2015)
14. Z. Zhang, M. Kamon, L. Daniel, Continuation-based pull-in and lift-off simulation algorithms for microelectromechanical devices. IEEE J. Microelectromech. Syst. **23**(5), 1084–1093 (2014)
15. A. Parent, A. Krust, G. Lorenz, T. Piirainen, A novel model order reduction approach for generating efficient nonlinear Verilog-A models of MEMS gyroscopes, in *ISISS* (2015)
16. M. Rewienski, J. White, A trajectory piecewise-linear approach to model order reduction and fast simulation of nonlinear circuits and micromachined devices. IEEE Trans. Comput. Aided Des. Integr. Circuits Syst. **22**(2), 155–170 (2003)
17. R. Zurawski, in *Embedded Systems Handbook – Embedded Systems Design and Verification* (CRC Press, 2009)
18. F. Pêcheux, C. Lallement, A. Vachoux, VHDL-AMS and Verilog-AMS as alternative hardware description languages for efficient modeling of multidiscipline systems. IEEE Trans. Comput. Aided Des. Integr. Circuits Syst. **24**(2), 204–225 (2005)
19. M. Radetzki, *Languages for Embedded Systems and Their Applications*, vol. 36 (Springer Netherlands, Dordrecht, 2009)
20. Accelera Systems Initiative, IEEE Standard SystemC – Language Reference Manual, IEEE Computer Standards Committee Automation Design, 2002
21. Accelera Systems Initiative, Standard SystemC AMS extensions – Language Reference Manual, 2010
22. A. Jantsch, *Modeling Embedded Systems and SoC's Concurrency and Time in Models of Computation* (Morgan Kaufmann, San Francisco, 2004)
23. P. Hartmann, P. Reinkemeier, A. Rettberg, W. Nebel, Modelling control systems in SystemC AMS – benefits and limitations, in *IEEE International SOC Conference (SOCC)* (2009)
24. T. Maehne, Z. Wang, B. Vernay, L. Andrade, C. Ben Aoun, J. Chaput, M. Louërat, F. Pêcheux, A. Krust, G. Schröpfer, M. Barnasconi, K. Einwich, F. Cenni, O. Guillaume, UVM-SystemC-AMS based framework for the correct by construction design of MEMS in their real heterogeneous application context, in *International Conference on Electronics Circuits and Systems (ICECS)* (Marseille, France, 2014)
25. L. Andrade, T. Maehne, A. Vachoux, C. Ben Aoun, F. Pêcheux, M.-M. Louërat, Pre-simulation symbolic analysis of synchronization issues between discrete event and timed data flow models of computation, in *Design, Automation & Test in Europe Conference and Exhibition (DATE)* (Grenoble, France, 2015)
26. D. Broman, C. Brooks, L. Greenberg, E.A. Lee, M. Masin, S. Tripakis, M. Wetter, Determinate composition of FMUs for co-simulation, in *Proceedings of the Eleventh ACM International Conference on Embedded Software* (2013)
27. G. Lasnier, J. Cardoso, P. Siron, C. Pagetti, P. Derler, in *Distributed Simulation of Heterogeneous and Real-Time Systems* (2013)

# Chapter 7
# Modeling and Simulation of the Power Flow in Smart Systems

**Sara Vinco, Alessandro Sassone, Massimo Poncino, Enrico Macii, Giuliana Gangemi, and Roberto Canegallo**

## 7.1 Introduction

Besides their heterogeneity in the type of devices, the other distinctive feature of smart systems is their being energy-autonomous; they are normally equipped with one or more devices able to harvest power from the environment, and, in many cases, also with elements that can smooth fluctuations in the harvested energy and deliver it to the various loads with a predictable "quality of service."

From the modeling and simulation point of view, the presence of these types of devices (generating or storing power) poses significant challenges. As a matter of fact, these components do not have a true functionality that can be described in the same way as a typical "functional" device like a processor core or an accelerator.

Let us consider for instance a scenario in which we are executing the RTL simulation of a processor core; functional simulation will track the cycle-accurate evolution of the core. If we want to monitor the power consumed by the core, we will leverage a properly characterized power model that will depend on the core state, frequency, and voltage so that we can track power cycle-by-cycle.

Now consider the case of a battery. It is evident that we cannot incorporate a battery to the RTL simulation scenario because the battery does not have any

S. Vinco (✉) • A. Sassone • M. Poncino • E. Macii
Politecnico di Torino, Torino, Italy
e-mail: sara.vinco@polito.it; alessandro.sassone@polito.it; massimo.poncino@polito.it; enrico.macii@polito.it

G. Gangemi
STMicroelectronics, Catania, Italy
e-mail: giuliana.gangemi@st.com

R. Canegallo
STMicroelectronics, Bologna, Italy
e-mail: roberto.canegallo@st.com

 
N. Bombieri et al. (eds.), *Smart Systems Integration and Simulation*,
DOI 10.1007/978-3-319-27392-1_7

equivalent of a logic-level functionality. The "functionality" of the battery has a completely different semantics and will involve how it delivers power (i.e., a requested current at a given supply voltage) over time based on the dynamics of the request. Therefore, the integration of energy storage devices (ESDs) into simulation implies speaking a "different language": it is not digital signals bringing information about functionality to be simulated, but rather power signals, i.e., voltage and current. The same clearly applies to devices generating power.

We could call this strategy a "native" power simulation, because it is power which is explicitly simulated as a (non digital) signal. Clearly, the link to functionality should not be lost. What components do in terms of computation is reflected by the power consumption.

Solving this issue is in principle relatively straightforward: one could for instance build an electrical circuit equivalent of the various components and simulate the whole thing using an electrical simulator like Spice. There are however two main difficulties with this strategy. Firstly, it could be tricky to abstract the behavior of some components as an electrical network; for instance, it is not obvious how a processor could be represented as a circuit element. Secondly, if the objective of simulation becomes to evaluate the power flow in a system over the typical duration of a battery discharge (several hours), it becomes unfeasible to use circuit-level simulators for that purpose.

A desirable solution should (1) provide a sufficiently high-level of abstraction in the simulation so to enable long simulation times, (2) abstract behavior of functional blocks to enable focusing on power as the quantity explicitly tracked in simulation, and (3) keep the link with functional simulation. As an additional desirable property, the simulation framework should be open and help re-using already existing models for the functional components.

This chapter describes a methodology for addressing the above issues based on this "native" power modeling and simulation approach, which relies on the use of a standard hardware description language (HDL), namely SystemC, and exploits its extension to model and simulate analog and mixed-signal components, thus enabling the concurrent simulation of functionality and the tracking of power. The approach relies on the definition of a unified interface for the various components of a smart system from the power perspective (ESDs, power sources, power converters, interconnects, and functional components—regarded as "loads") and a "meta-modeling" approach in which models of actual devices are fitted to the model templates defined by those interfaces.

The effectiveness of the proposed approach is demonstrated on an industry-strength smart system prototype, where significant performance speedups against model-based tools were achieved.

## 7.2 Power View of a Smart System

Figure 7.1a depicts the classical architecture of a smart system, consisting of a bunch of sensors generating signals that are first amplified and converted in the digital domain. Here, the digitized signal is variously processed (e.g., filtered) by custom and/or programmable components (generically labeled "digital processing" in the figure). The results are reconverted into the analog domain and returned to the user in the form of some physical signals through appropriate actuators. Superimposed to this "functional flow" (i.e., involving how the data are processed), the dotted arrows denote another flow of signals, this time involving the distribution of power. Similarly to the functional flow, but involving different components, the power is extracted from the environment by a set of scavengers exploiting a multitude of transduction mechanisms from various domains (mechanical, electromagnetic, chemical, acoustic, and thermal) into the electrical one. This harvested power is delivered either to the various functional components, which will *consume* it, or to devices that are able to *store* it.

From the power perspective the view of the smart system can be refined to that of Fig. 7.1b, in which the behavior of the functional components is now immaterial—

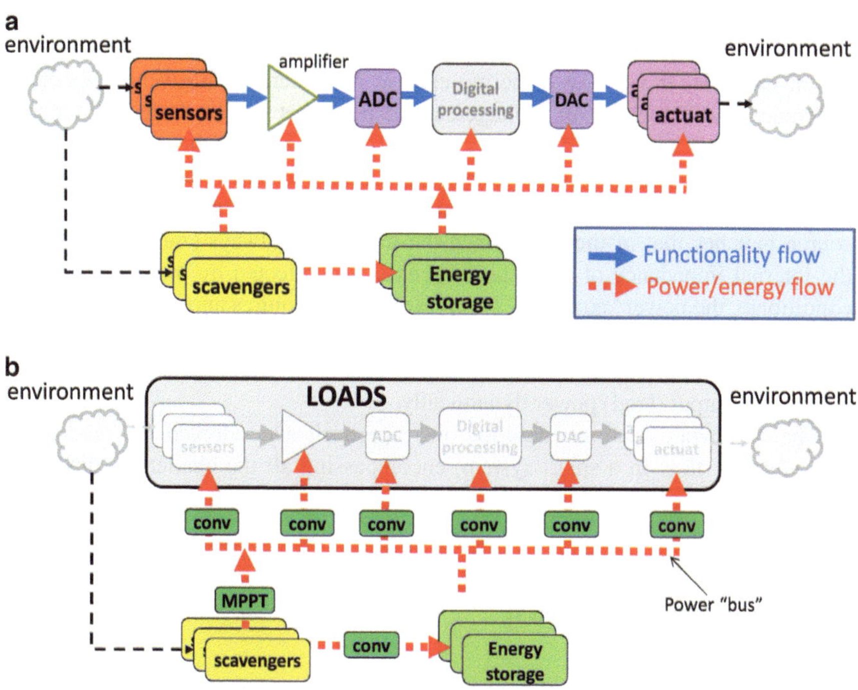


**Fig. 7.1** Functional and power "Flows" in smart system. A smart system from the functional perspective (**a**) and from the power perspective (**b**)

what matters is that this behavior translates into a consumption of power over time. For this reason all these components have the same "behavior" from the power perspective and are seen as *loads*.

Figure 7.1b emphasizes another characteristic that translates into a technical difficulty while modeling and simulating the flow of power. Power represents the current drawn or generated by a component at some voltage level; it is evident that it is not possible to directly interface components that are at different voltage levels. Appropriate voltage conversion is therefore needed (green blocks labeled **Conv** in the figure) between any two blocks. Notice that the various functional components often have disparate supply voltage levels (e.g., digital, analog, I/Os). This is typically implemented by converters (mostly DC/DC, but also AC/DC for alternate current power sources).

In order to allow scalability, similarly to what is done to "functionally" connect components of an architecture, it is preferable to avoid point-to-point connections and rather use a "power bus," which represents a common connector at a given voltage. We will call this the DC-bus, since the majority of components involved in a smart system uses direct current. While in some application domains (e.g., automotive, microgrids) such buses are often standardized, this is not the case for electronic systems.

In the case of the power sources (the scavengers) there is an additional complication. These elements typically generate variable amounts of power (for a given value of the harvested quantity) that depend on the electrical connection with the load(s). As an example, if we take a thermo-electric harvester, for a given temperature difference, the actual power (voltage/current pair) that can be extracted depends on the matching with the load. This is the well-known issue of the *maximum power transfer*, which has a very simple electrical solution in the *maximum power theorem*, stating that the extracted power is maximum when the internal impedance of the source matches that of the load. Since the load is not always constant (and so is the internal impedance of the source because it is affected by the environmental conditions), the power "adaption" of a power source requires appropriate circuits to **dynamically** extract the maximum power; these are called maximum power point transfer (MPPT) circuits, and are variants of basic converters that can match input (source) to output (load) power dynamically.

The following sections describe the approach we adopted to model the power/energy flow in a smart system, encompassing all the challenges described above.

## 7.3 Vision and Approach

The goal of the simulation of the power perspective of smart systems is to *trace power flows*, with the goal of achieving an early assessment of system behavior and of validating the dimensioning of the energy providers. In the proposed framework, power flows are represented in terms of voltage levels and current demand/production over time of the various components. For this reason, voltage (`V`) and current (`I`) are the main dimensions of each component, and will be the focus of overall simulation.

Components naturally have *different roles w.r.t. the power flow*, i.e., each system includes component that either consume, generate, distribute, or store energy. This difference is reflected upon simulation: components with different roles will have different interfaces and models. It is thus necessary to provide a classification of system components and to construct a template to be used as reference for the overall framework, as presented in Sect. 7.3.1. Section 7.3.2 completes the template with the interface of each class of components. Finally, Sect. 7.3.3 outlines the main models available in the literature, together with some implementation guidelines for the construction of the overall framework.

### 7.3.1 System Architectural Template

Figure 7.2 outlines the proposed template of a smart system from the power perspective. Each system features a certain number of *loads*, i.e., components that require a given amount of power to implement a certain functionality (e.g., digital cores, MEMS, analog, and RF components). Power can be provided by either

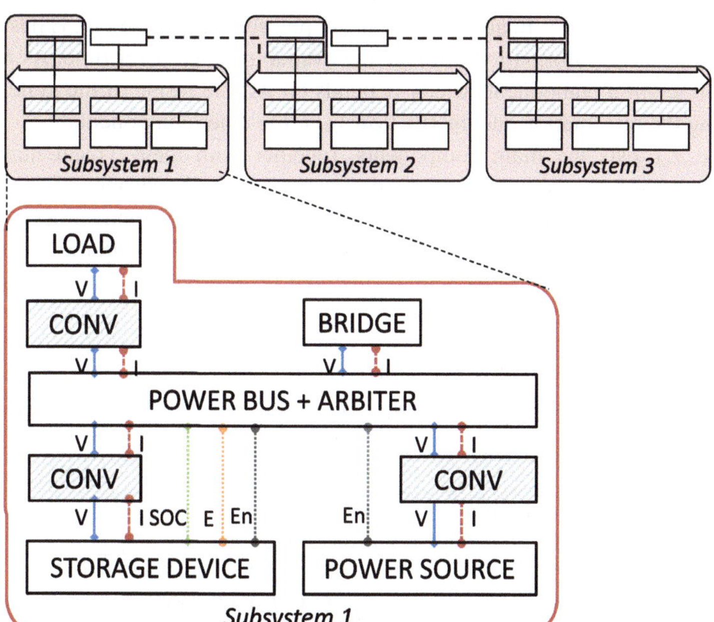


**Fig. 7.2** Template of the reference architecture of a smart system from the power perspective, including the power bus and the connected components

ESDs or power sources. *ESDs* can be of different natures, ranging from batteries to supercapacitors and fuel cells. *Power sources* are almost infinite sources of power, such as photovoltaic cells or thermo-electric energy generators, used for either satisfying the loads power demand or for charging ESDs. ESDs and power sources are managed by an *arbiter*. The arbiter monitors the state of charge (SOC) of the ESDs, to determine whether they can provide energy or they have to be charged. Furthermore, the arbiter determines what ESD or power source to use, based on the loads request for power. All components are connected through a *power bus*, that allows for the energy to combine and propagate within the system (either as an ideal conductor or with some power loss). Each component is connected to the power bus through a *converter* module, necessary to maintain compatibility of voltage levels. Connections between different subsystems are finally managed by *bridges*, that behave as converters between two power buses.

It is important to note that any system may contain a number of each type of components. The fundamental requirement is that any system contains at least one ESD or one power source, to provide energy to load components.

### 7.3.2 Power Interfaces

The different role of components impacts on their simulation characteristics. Energy providers may be enabled or disabled, depending on the operating conditions (e.g., to prefer a power source to an ESD, whenever possible). Furthermore, ESDs require an intrinsic status information, to trace the available charge over time. As a result, the interface of a component strictly depends on its role inside of the system, thus reflecting the information and energy flow *w.r.t.* the other components.

Table 7.1 lists the main components, together with their typical number of occurrences and the ports and connections modeled in the system. (For the sake of clarity, it is assumed that the system contains only one component per type, i.e., $l = s = p = 1$.) In this version of the system, the power bus and the arbiter have been merged in a single component, as the goal is to provide a simple simulatable

**Table 7.1** Interface of each class of components

| Component | Instances (#) | Power interface |
|---|---|---|
| Load | $l$ | $(V, I)$ |
| ESD | $s$ | $(V, I, \mathrm{SOC}, E, \mathrm{En})$ |
| Power source | $p$ | $(V, I, \mathrm{En})$ |
| Converter | $c = s + p + l$ | $(V, I, V, I)$ |
| Arbiter | 1 | $((\mathrm{SOC}, E, \mathrm{En})^s, (\mathrm{En})^p, (V, I)^c)$ |
| Power bus | 1 | $((\mathrm{SOC}, E, \mathrm{En})^s, (\mathrm{En})^p, (V, I)^c)$ |
| Bridge | $b$ | $(V, I, V, I)$ |

interface, rather than to reflect the physical components of the power system. This *enhanced power bus* constitutes an abstraction of the power behaviors of the overall heterogeneous system.

The interface of each component describes what power information is shared with the remainder of the system. $V$ and $I$ are voltage and current, respectively. SOC and $E$ are the SOC and the energy (i.e., capacity) of the ESD components. En is an enabling signal, used by the arbiter to activate an ESD or a power source. Environmental parameters that may influence the behavior of components (e.g., temperature and solar radiation) are not modeled on the component interface, as they affect the component behavior rather than the overall system.

The system energy flow moves energy from ESDs and power sources to loads. As a result, the interface of each class of components is as follows:

- *loads* share information about required current ($I$) and operating voltage ($V$);
- *ESDs* share their voltage ($V$), whereas they are provided with the current demand ($I$) of the system. Furthermore, they must communicate their SOC and their nominal capacity ($E$), so that they can be activated by the arbiter through an enable signal (En) depending on the actual energy capability;
- the interface of *power sources* includes the supplied current ($I$) and voltage ($V$), and an activation signal (En);
- the key role of the *arbiter* is to determine what ESD or power source to use, based on the loads request for power. Therefore, its interface includes a couple of (SOC, $E$) ports for each ESD and an activation signal (En) for each ESD and power source;
- the *power bus* connects all components, thus its interface consists of a number of $I$ and $V$ ports;
- *converters* feature a couple of ports ($I$, $V$) for the input and output values of current and voltages of each connected component;
- *bridges* act as converters between different power buses, and thus feature a couple of ports ($I$, $V$) for each connected bus.

### 7.3.3 *Power Models and Implementation*

The definition of the system template and the formalization of simulation interfaces constitute a skeleton of the simulation framework, as they formalize energy and information flows. Each component must then be implemented by adopting state-of-the-art models. An extensive presentation of the available models is out of the scope of this section, as they will be presented in Sect. 7.4. Nonetheless, it is necessary to discuss some of their characteristics to determine the desired features of the simulatable framework.

Available models strictly depend on the modeled class of components, but they can be classified in terms of abstraction level, i.e., in terms of accuracy *w.r.t.* the simulated physical phenomena. *Functional models* implement component evolution with a function (e.g., modeling an equation, a finite state machine, or even a simple waveform). A well-known example of functional model is the Peukert's model for batteries [19]. On the other hand, *circuit models* emulate the evolution of a component by building an equivalent electrical circuit [18], reproducing the internal dynamics of the component.

The proposed framework for simulating the power aspects can be implemented in a number of languages, ranging from MATLAB/Simulink up to HDLs, depending on the desired characteristics and tradeoff in terms of speed and accuracy. However, the coexistence of heterogeneous levels of detail imposes some requirements on the language adopted for its implementation. It must indeed *support different levels of abstraction*, to model both discrete-time functional models and circuit elements. Furthermore, the language must be *modular*, i.e., separate the modeling of component interface and connections from the implementation of its behavior. It must thus be possible to adopt different implementations for the same component, depending on the target (i.e., accuracy or performance), without affecting its interface and the remainder of the system. Finally, the adopted language should *enhance reuse*, to allow the construction of a library of configurable models that may be reused at later stages of design or in future systems.

## 7.4 Power Components

Models available in the literature strictly depend on the target class of components. This section sketches how each class can be modeled, by adopting effective state-of-the-art models suitable for the proposed system-level approach. *Note that the current work is restricted to direct current (DC) systems, as smart system components rarely support alternate current (AC). As an example, the only AC power sources are scavengers based on piezo-electric mechanisms; in this case we assume the generated AC power is rectified directly inside the power source device, which therefore outputs DC voltage and current.*

### *7.4.1 Loads*

The term "load" comprises any component requiring a given amount of power to implement a certain functionality, e.g., digital cores, MEMS, analog, and RF components. This chapter treats such components with a very different perspective *w.r.t.* the remainder of the book. Their functional evolution is indeed disregarded, and components are seen as *black boxes that require a certain amount of current (`I`) at a given voltage level (`V`)*, as shown in Fig. 7.3. An additional enabling signal (`En`) may be used to make the load controllable, i.e., to activate or deactivate its functionality depending on the power flows in the system. As an example, in case that the energy

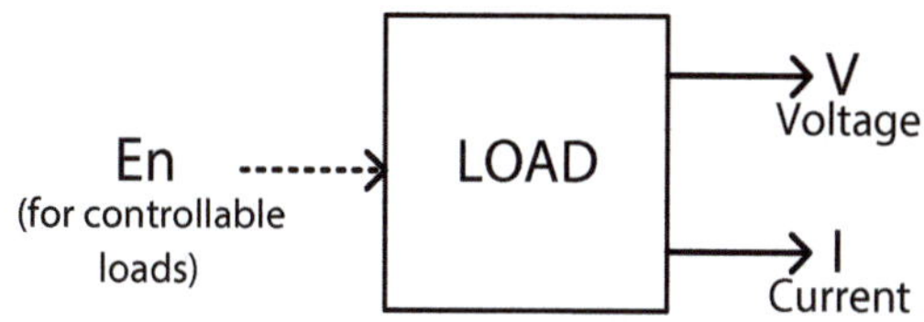


Fig. 7.3 Generic interface of a load

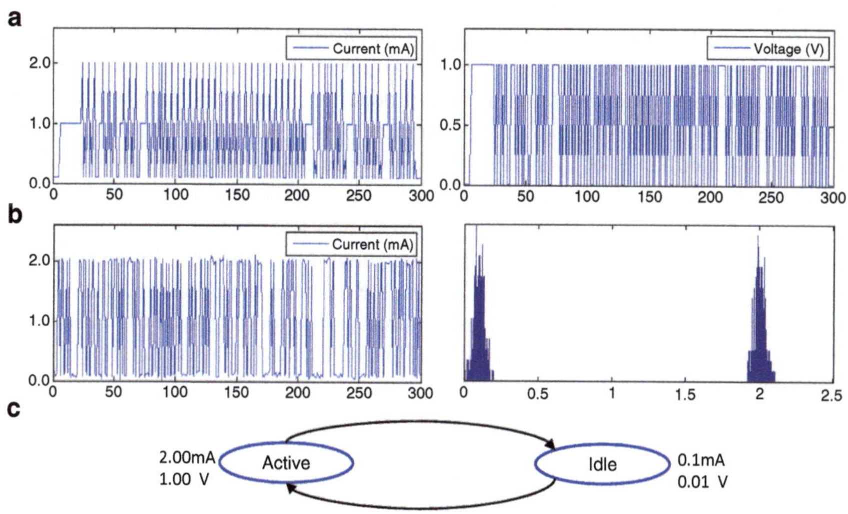


**Fig. 7.4** Examples of power models for loads: execution traces for current and voltage (**a**), synthetic statistic traces (**b**), and a power state machine (**c**)

production is extremely limited *w.r.t.* the demand, it may be necessary to disable non-critical loads, to guarantee correct operation of critical components.

The interface adopted for loads highlights that the focus of the overall chapter is solely on tracing the smart system energy flows, represented in terms of voltage levels and current demand/production over time of the various components. The main consequence of this view is that models for functional components are simplistic, to the point that they may fall back to synthetic `V` and `I` traces over time. Figure 7.4 shows some of the mostly adopted models for loads, divided depending on the different levels of adherence to the component evolution.

The most accurate models are *execution traces*, obtained with experimental measurements applied to the component during a typical excerpt of its execution. These models are made up of a couple of waveforms reproducing how current demand and voltage actually evolved over time during the sampled execution excerpt. The traces may be repeated periodically, to simulate longer executions. Figure 7.4a exemplifies this by showing a trace for current demand (left) and one for voltage level of the component (right).

In case experimental measurements are not possible or they are considered too accurate for simulation purposes, they can be replaced with *synthetic traces*

(Fig. 7.4b). The accuracy of such models strictly depends on their construction process. Typical consumption and voltage values can be extracted from component datasheet, and the trace may be built by relying on statistical information, by reflecting some “typical” consumption profile of the component, e.g., power values that are more frequent than others. Figure 7.4b shows an example trace modeled as a bimodal distribution (left), where the most frequent values correspond to the typical active and idle current demands (right).

Finally, loads may be modeled as *state machines* listing the internal states of the component (Fig. 7.4c) [5]. Transition from one state to another may depend on overall system information or on timers, reproducing a typical execution flow of the component and its dynamic power management policy.

The presented kinds of models stress even more that, when focusing on power, the dependency *w.r.t.* functional execution is broken. The focus is indeed on the dynamics of the energy providers (e.g., batteries and scavengers), rather than on the actual load behavior. In case the adopted language for framework implementation supports functional modeling, it is still possible to make the power models dependent on the functional execution, e.g., by predicating the edges of the finite state machines with conditions on functional evolution. However, this is out of the scope of this chapter.

### *7.4.2 Power Sources*

Among the various components, the most diverse of all are power sources, i.e., elements that generate power by transforming some environmental quantity in electrical energy. The variety of their characteristics tends in fact to follow the scale of the relative system: they range from the μW/mW scale of MEMS-based energy micro-scavengers to the MW scale of large wind turbines, and the very scavenging mechanism can be quite different.

This variety in the typologies of power sources makes their modeling poorly scalable and marginally re-usable, thus complicating the objective of using an as much as possible unified modeling approach. The large variety of types of power sources is reflected thus by the many available options for their modeling, ranging from multiphysics-based mechanical models [8, 9] and equation-based mathematical models [11], up to electrical circuit equivalent models [4, 22], and functional (continuous or discrete-time) macro-models suitable for system-level simulation [10, 31].

Any power source can be considered as a component that generates *voltage and current waveforms over time*, as shown in Fig. 7.5 (signals `V` and `I`). The power produced by the power source strictly depends on the harvested quantity, modeled as an input waveform over time (signal `H`). The power source may be further affected by other environment characteristics, such as temperature (signal `E`). The proposed model is clearly agnostic both of the type of power source and of the scale of managed energy.

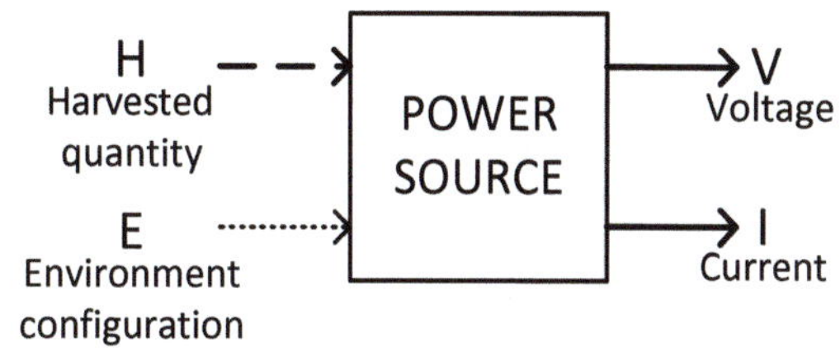


**Fig. 7.5** Generic interface for power sources

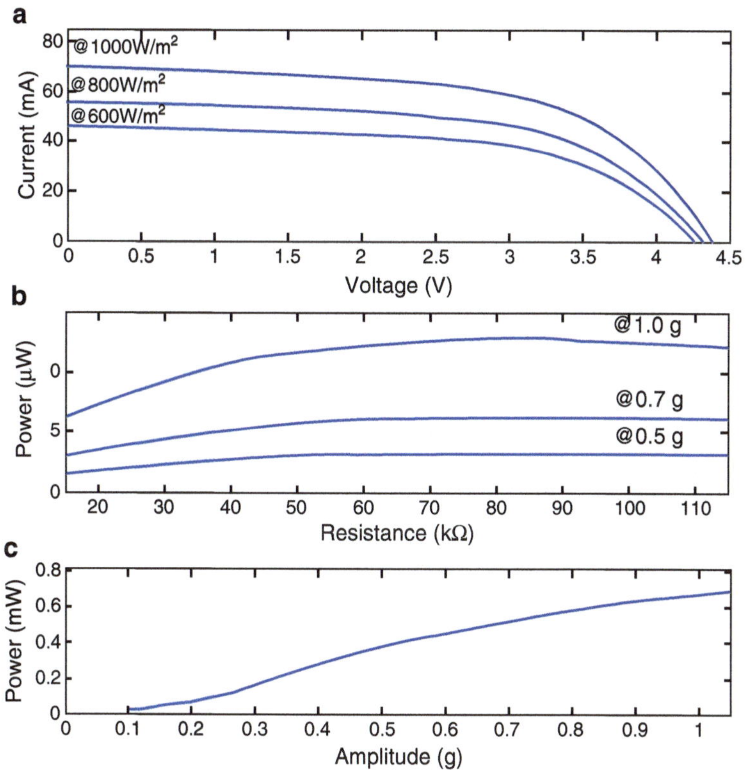


**Fig. 7.6** Examples of datasheets graphs: (**a**) a Class 1 graph for the photovoltaic cell in [20] (each curve is associated to a value for irradiance); (**b**) a Class 2 graph for the piezo-electric harvester in [3] (parameterized *w.r.t.* the corresponding acceleration value); (**c**) a Class 3 graph for the piezo-electric harvester in [15]

The analysis of more than 50 datasheets for different types of power sources, including photovoltaic cells and piezo-electric harvesters, leads to the definition of *three main templates* for describing power source behavior:

- Class 1 graphs are *Current* vs. *Voltage or Power* vs. *Voltage* graphs (Fig. 7.6a), that reproduce the dependency *w.r.t.* the harvested quantity through a number of current/voltage (power/voltage) curves, each one associated with a specific value for the harvested quantity [6, 15, 23]. Power/voltage plots are similar, since they are straightforwardly derived from current/voltage ones by multiplying voltage and current values.

- Class 2 graphs are *Power* vs. *Resistance* graphs. This type of specifications, often used for piezo-electric power sources [3], models the power source as a family of power/resistance curves, each associated with a specific value of the harvested quantity (Fig. 7.6b).
- Class 3 graphs are *Power* vs. *Harvested Quantity* graphs, popular as specifications of piezo-electric harvesters [15] (Fig. 7.6c). Here voltage is defined in the datasheet as one or more pre-defined voltage levels, while current must be derived from the voltage and the power behavior.

As a reference form, we adopt Class 3, i.e., *power* vs. *harvested quantity* graphs, as these curves conceptually represent the actual "behavior" of a power source, i.e., a device that generates power according to some environmental quantity. It is thus necessary to re-cast the other two classes of graphs to the canonical form, by making the dependency on the harvested quantity explicit.

Both the classes associate *a curve* to each value of the harvested quantity, thus not univocally identifying an output quantity (i.e., either power or voltage). The necessary transformations reduce the graph to the canonical form by extracting the *maximum power point* (MPP) of the device *w.r.t.* the harvested quantity [12], i.e., by determining voltage and current values that yield the maximum power for a given environmental condition. In case of power vs. voltage graphs (Class 1) or power vs. resistance graphs, the MPP for each value of the harvested quantity is the maximum of the corresponding curve. Current vs. voltage graphs are instead reduced to power vs. voltage graphs. The power source general model can then be easily constructed by plotting the MPPs for the known values of the harvested quantity, which can be interpolated to define a continuous curve. Figure 7.7a highlights the MPP for each irradiance value on the current vs. voltage Class 1 graph in Fig. 7.6a (left) and the resulting canonical form (right-hand side). Figure 7.7b shows the result of applying the same approach to the Class 2 graph in Fig. 7.6b.

The behavior of the power source is then described as a pair of curves extracted from the canonical form, modeling voltage and power vs. the harvested quantity. Note that the extraction of such information is agnostic both of the type of power source and of the scale of managed energy, and it allows to include the power source models in a wide range of simulation infrastructures.

### *7.4.3 Energy Storage Devices*

Although the approach described hereafter conceptually applies to any device capable of storing energy, in this chapter we restrict ourselves to ESDs that are typical of smart systems, namely, *batteries* and *supercapacitors*.

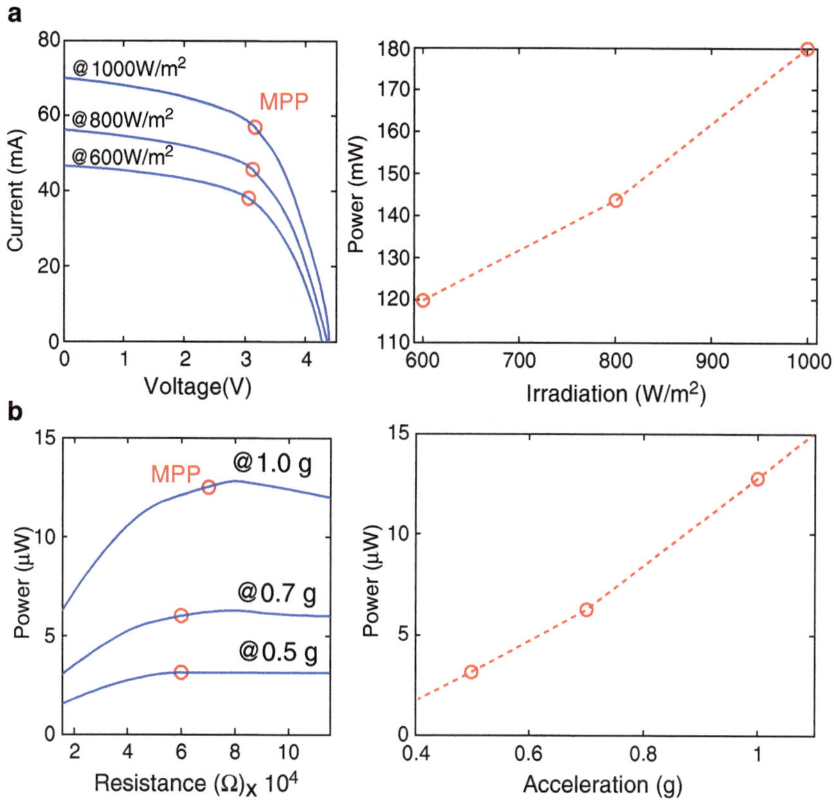


**Fig. 7.7** Construction of the generic power source model for the photovoltaic cell in Fig. 7.6a (Class 1) and of the piezo-electric harvester in Fig. 7.6b (Class 2)

**Fig. 7.8** Generic interface of an energy storage device (ESD)

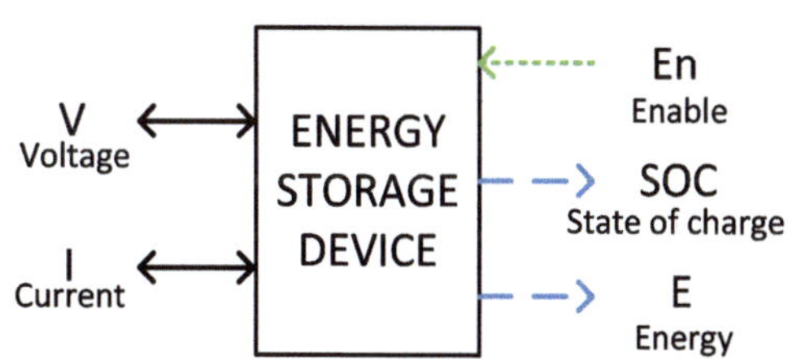


As introduced in Sect. 7.3.2, one key feature of our methodology is to assume a unique interface for a given type of component. Similarly to power sources, ESDs require a general interface that can fit the large variety of available devices. As described in Sect. 7.3.2, such interface is as depicted in Fig. 7.8.

The interface contains the two "native" quantities, i.e., voltage (`V`) and current (`I`); notice that unlike loads and power sources, for ESDs these signals are bidirectional, since it is expected that an ESDs can both accumulate and deliver energy.

Moreover, an ESD features two state signals, denoting respectively its `SOC` and its residual nominal capacity (`E`). Although the two are related, their relation depends on the type of ESD and therefore two separate signals are maintained. Finally, a control signal `Enable` allows one to selectively disconnect the ESD from a load or source to implement specific management policy.

#### 7.4.3.1 Modeling of Batteries

We consider two popular models with different tradeoffs between accuracy and complexity, namely a functional model based on Peukert's law and a behavioral model based on a circuit equivalent of a battery. These two models are general enough to be applied to *any* type of battery (chemistry or form factor) and they can easily be identified based on a limited set of information typically available in battery datasheets [18].

The functional model based on *Peukert's equation* expresses the non-linear relation between the battery current $I$ and the equivalent capacity (represented in our model by signal $E$) through the Peukert's coefficient $n > 1$ [19]:

$$E = I^n \cdot t$$

here $t$ denotes time. The SOC of the battery in this model is obtained as

$$\mathrm{SOC} = 1 - \frac{I \cdot t}{E}$$

This model is able to track the load-dependent capacity property of a battery, but not its sensitivity to the current dynamics; current $I$ is assumed to be constant. Identification of this model, i.e., determining its only parameter ($n$), can be easily done by tabulating (current, discharge times) pairs from the typical battery discharge curves contained in datasheets (Fig. 7.9a) and by fitting these values against a $\frac{1}{I^n}$ curve equation (Fig. 7.9b) [18].

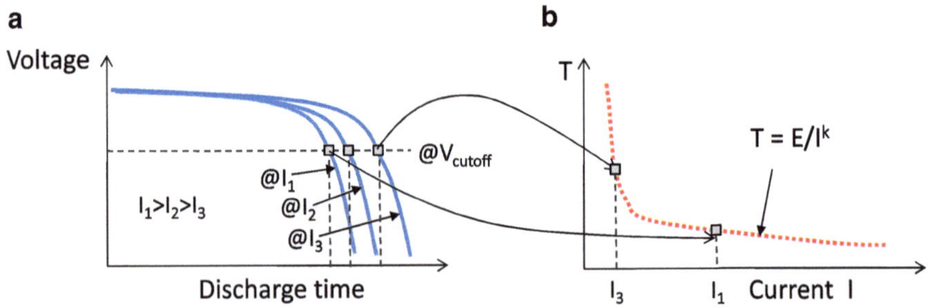


**Fig. 7.9** Identifying Peukert's equation from battery discharge curves. Battery discharge curves (**a**) and corresponding Peukert's equation built from the extracted (current, discharge time) pairs (**b**)

The second and more accurate model is an *electrical circuit equivalent* that mimics the battery behavior [21]. Although many circuit equivalent have been proposed in the literature, we adopt a relatively simple model (Fig. 7.10) that simplifies its identification process by requiring only a few, publicly available information.

The model consists of a left part that models the battery lifetime, and consists of a capacitor $C$ (representing the storage capacity of the battery in Amp-hours), and a current generator representing the current requested by the load $I_{batt}$. The rightmost part models the transient behavior (dynamics) and is mainly characterized by the resistor $R$ that denotes the battery internal resistance; the voltage drop $R \cdot I_{batt}$ that affects the actual battery output $V_{batt}$ mimicks the fact that the battery voltage is adversely affected by larger currents. The connection between the two parts is modeled by the voltage generator $V_{OC}(V_{SOC})$: it represents the fact that the open-circuit (OC) voltage of the battery depends on its state-of-charge (SOC), represented by the potential $V_{SOC}$ of the capacitance. Notice also that in the most general case, the internal resistance is also depending on the SOC.

This baseline model is quite modular and can be augmented by incorporating additional elements such as [18]:

- Accounting for the frequency dependence of the internal resistance, i.e., making it an internal impedance; this corresponds to adding one or more RC blocks with different time constants to model short- or long-term components of the transient behavior.
- Accounting for the self-discharge of the battery; this is done by adding a resistor in parallel to the capacitance $C$.

Other "non-behavioral" effects such as temperature dependence or aging can also be included by adding electrical components in the circuit equivalent; since these variants represent long-term effects that are noticeable only across several charge–discharge cycles (days or months) they are not considered in this chapter, and we refer the reader to [7] for a comprehensive overview.

The basic model of Fig. 7.10 can be easily identified by using the (voltage, capacity) or (voltage, SOC) curves provided in most datasheets. At least two curves (for different discharge currents) are needed to determine the internal resistance $R$ (Fig. 7.11) as described in [18]. At a given voltage, the two curves provide two reference points of the battery voltage $V_{batt}(SOC_1)$ (corresponding to a current $I_1$)

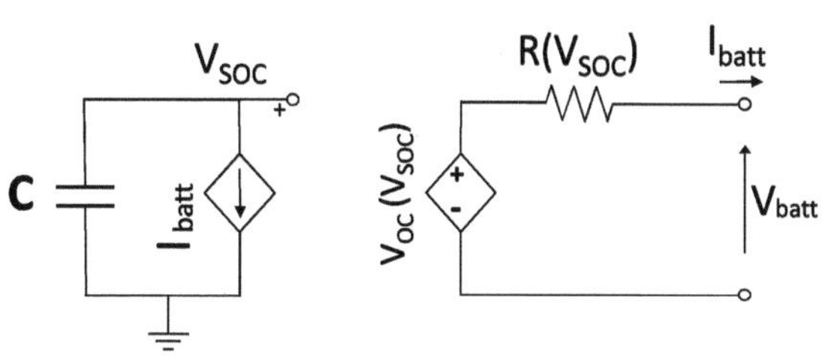


**Fig. 7.10** Circuit-equivalent model template used for batteries

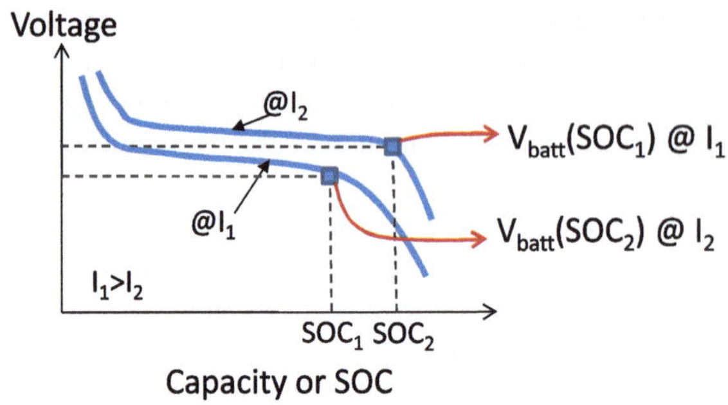


**Fig. 7.11** Model parameter identification for the model of Fig. 7.10

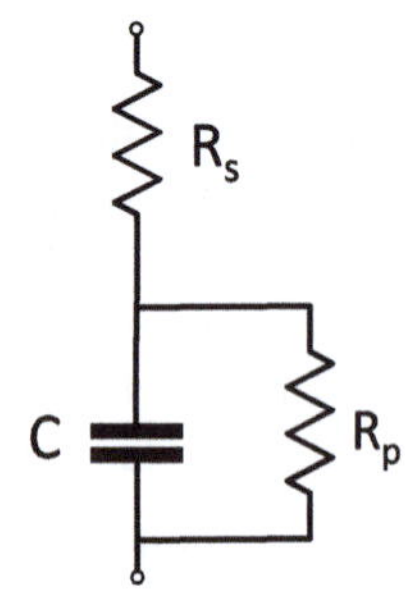


**Fig. 7.12** Equivalent circuit model of a supercapacitor

and $V_{\mathrm{batt}}(\mathrm{SOC}_2)$ (current $I_2$). By writing the voltage equations in the right-side mesh of the circuit we can write two equations (corresponding to the two currents $I_1$ and $I_2$) that allows determining the two unknowns, i.e., $R(V_{\mathrm{SOC}})$ and $V_{\mathrm{OC}}(\mathrm{SOC})$.

#### 7.4.3.2 Modeling of Supercapacitors

Being an electrical device, a well-accepted model for supercapacitors is an equivalent electrical circuits; as for batteries, the circuit can have different levels of complexity, depending on the phenomena considered in the model [13].

The circuit considered in this chapter is depicted in Fig. 7.12. This first-order model consists of three main components. The capacitor obviously represents the nominal capacitance of the supercapacitor. The series resistance $R_s$, usually referred to as the *equivalent series resistance* (ESR), is the main contributor to power loss during charging and discharging of the capacitor; its values are generally quite small (in the order of m$\Omega$s). The model includes also a parallel resistance $R_p$ which affects s the self-discharge of the capacitor. $R_p$ is always much larger than the ESR (order of k$\Omega$s).

More sophisticated supercapacitors models have been devised in the literature; they might include a series inductance, a voltage-dependent capacitor in parallel with $C$ [33], or replace the $(C, R_s)$ pair with a series of cascaded RC elements to model the frequency dependence of the capacitance. Although the model Fig. 7.12 abstracts away these second-order effects, it is accurate enough for a system-level exploration of the power flow for our context.

Identification of the three main model parameters ($C$, $R_s$, and $R_p$) in Fig. 7.12 is immediate for the former two, but less obvious for $R_p$. Supercapacitor datasheets systematically provide data for $C$ and $R_s$, but very seldom information about $R_p$. For this reason, in this chapter we disregard $R_p$ in the model (i.e., assume $R_p = \infty$) and stick to a simple R-C series model of the supercapacitor. In this way, the two values of $C$ and $R_s$ need simply to be transcribed from the corresponding datasheet.

With this choice, we neglect the self-discharge behavior of the supercap; as a matter of fact, the battery models considered in the previous section do not include the modeling of self-discharge either (although this effect is less evident in batteries). In that respect, therefore, the battery and supercapacitor models used in our analysis are consistent in terms of modeled effects.

### 7.4.4 Power Conversion and Steering

*This section encapsulates two classes of components, both in charge of steering the flow of power inside of a system. The* power bus *provides a reference voltage level to all system components.* Converters *allow to maintain compatibility of voltage levels whenever components operating at different voltages are connected. Note that bridges are viewed as converters between special components, i.e., power buses. Thus, they are not presented separately.*

#### 7.4.4.1 Power Bus

The power bus, also called DC bus or CTI (charge transfer interconnect), is essential to carry around a given voltage level similarly to what happens to "functional" buses in a computing system.

Generally speaking, the power bus is nothing but a wire that holds a given voltage level. Therefore, any model suitable for electrical interconnects can work to model the power bus—options range from ideal wires to distributed RLC interconnects.

Accurate models of the power bus are particularly relevant in larger-scale energy systems such as micro-grids or hybrid electric vehicles, where interconnections connect components over large areas and support high voltages and can thus incur in significant losses. In smart systems, where the lengths of this interconnect is relatively small and voltages in the order of a few volts are supported, we do not need an extremely accurate model for the power bus. In our work, we envision two simple options with increasing accuracy:

1. an *ideal dc bus* consisting of an ideal wire and a voltage generator at the desired bus voltage;
2. a *purely capacitive bus*, whose capacitance is determined based on the overall bus length and size.

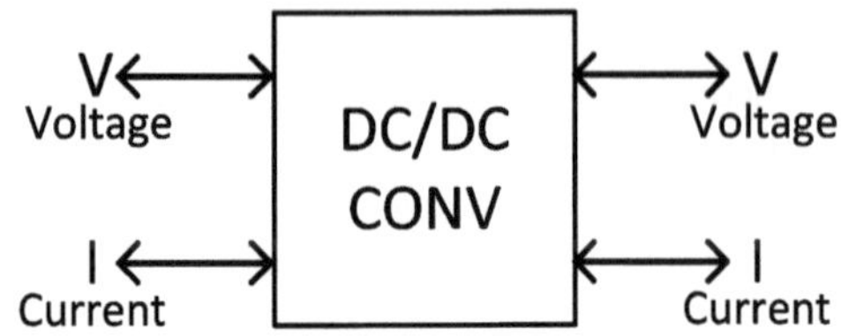


**Fig. 7.13** Generic interface of a DC/DC converter

#### 7.4.4.2 Power Converters

As mentioned in Sect. 7.2, interfacing two different voltage domains requires some adaptation. Since in our context all signals are DC, only DC/DC conversion is relevant. The generic interface of a DC/DC converter (Fig. 7.13) consists of two pairs of voltage and current signals. For generality, in the figure signals are bi-directional to denote the possibility of a bi-directional flow of the power.

Functionally speaking, the DC/DC converter simply adapts input power to match output power by mean of appropriate circuitry. This process can be characterized by the efficiency of the conversion $\eta$:

$$\eta = \frac{P_{\text{out}}}{P_{\text{in}}} = \frac{V_{\text{out}} I_{\text{out}}}{V_{\text{in}} I_{\text{in}}}$$

The conversion is in general non-ideal and not all input power is transferred to the output, hence $\eta < 1$. The difference between $P_{\text{out}} - P_{\text{in}}$ represents the *losses* of the converter. A detailed description of architectural and circuit variants of DC/DC converters, as well as a characterization of the losses is out of the scope of this chapter; for a detailed overview the reader can refer to [32].

Since the DC–DC converter is an electronic device, in principle a circuit-level model consisting of the interconnection of the discrete components would guarantee the highest accuracy. However, this would require a specific model for any specific type of converter (e.g., switching vs. linear) and would slowdown the overall simulation. Conversely, a system-level, functional model of the DC–DC converter describes its efficiency does capture all the non-idealities and is also independent of its implementation.

Key for such a model is the determination of the model parameters, i.e., which quantities affect efficiency. In general, conversion efficiency is affected, in order of relevance, by (1) output current $I_{\text{out}}$, (2) difference between input and output voltage $\Delta V = |V_{\text{in}} - V_{\text{out}}|$, and (3) absolute values of $V_{\text{in}}$ and $V_{\text{out}}$. This applies to the most complex converter architectures containing diodes, inductors, and capacitors (*switching* converters). For simpler architectures like those based on resistive elements (*linear* converters), efficiency is essentially determined by $\Delta V$.

Figure 7.14 shows an example efficiency curve from a step-down switching converter by Maxim [14] in which we can notice how efficiency is quite far from the ideal value, especially when $I_{\text{out}}$ gets smaller.

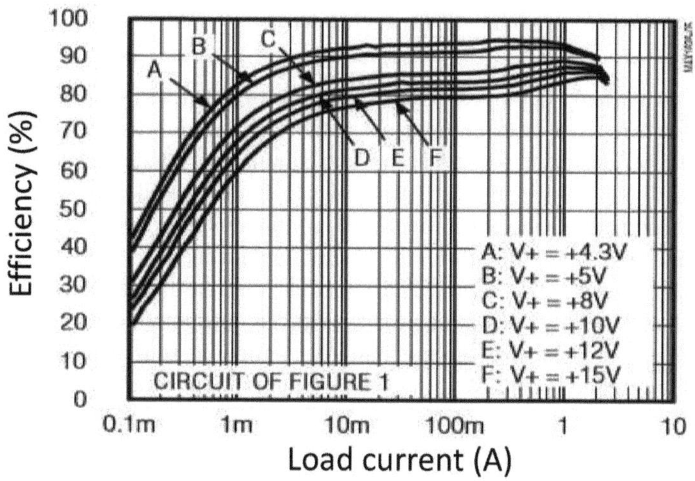


**Fig. 7.14** Converter efficiency curves for the maxim 1626 step-down converter (when assuming $V_{out}$ equal to +3.3 V)

Based on these considerations, we model the power flow of a DC–DC converter through its efficiency, as a quadratic function of the two most relevant parameters ($I_{out}$ and $\Delta V$), as follows:

$$\eta(I_{out}, \Delta V) = k_1 I_{out}^2 + k_2 \Delta V^2 + k_3 I_{out} + k_4 \Delta V + k_5.$$

In case the dependency of efficiency on one of the two parameters is irrelevant (e.g., $I_{out}$ for linear converters), that parameter will simply not appear in the equation. Coefficients $k_i, i = 1, \ldots, 5$ are calculated by least-square fitting of the model template, using data obtained by digitizing the efficiency curves provided in the datasheet of the specific device.

## 7.5 Implementation

The system-level vision of the power dimension of a smart system proposed in Sect. 7.3 may be implemented in a number of languages and frameworks, depending on the desired speed/accuracy tradeoff and on the need for integration with other simulation tools. In this work the choice fell on SystemC and on its Analog Mixed Signal (AMS) extension [1, 2] for two main motivations. First of all, SystemC is a standard language, thus extensible and free from compatibility and reuse issues, typical of proprietary tool. Furthermore, both SystemC and SystemC-AMS cover a number of abstraction levels, thus allowing to cover a wide range of models and to find appropriate simulation speed/accuracy tradeoffs. The following of this section details the implementation process.

### 7.5.1 SystemC and Its AMS Extension

SystemC is a widely deployed extension to C/C++ for describing HW constructs, ranging from register-transfer level up to transactional level [1]. Its recent AMS extension was designed for modeling and simulating interacting analog/mixed-signal functional subsystems [2]. This allows to extend the adoption of a SystemC-based environment also to extra-functional, continuous time domains [30].

SystemC-AMS provides different abstraction levels to cover a wide variety of domains. *Timed Data-Flow* (TDF) models discrete-time processes, that are scheduled statically by considering their producer–consumer dependencies. *Linear Signal Flow* (LSF) supports the modeling of continuous time behaviors through a library of pre-defined primitive modules (e.g., integration, or delay), each associated with a linear equation. *Electrical Linear Network* (ELN) models electrical networks through the instantiation of pre-defined primitives, e.g., resistors or capacitors, where each primitive is associated with an electrical equation. In case of ELN or LSF descriptions, a SystemC-AMS AD solver analyzes the ELN and LSF components to derive the equations modeling system behavior, that are solved to determine system state at any simulation time.

### 7.5.2 Power Simulation with SystemC-AMS

The implementation of the power perspective of a smart system requires the simultaneous simulation of models at different levels of abstraction, e.g., waveforms and circuit-equivalent models. This may impact on synchronization: low level models may introduce too many synchronization points, and it would thus be difficult to determine a correct synchronization mechanism with higher level models. To this extent, we adopted the TDF level of abstraction for all interfaces and connections. This implies that synchronization between components happens at pre-defined fixed time steps, e.g., once every millisecond (of simulated time). On the contrary, component models can be implemented with the most suitable level of abstraction, ranging from SystemC TLM up to SystemC-AMS TDF, or ELN itself. This allows to determine a good tradeoff between accuracy of the model and effectiveness of the synchronization mechanism.

To this extent, components are implemented as SystemC modules (i.e., instances of `SC_MODULE`), to leave freedom to adopt any level of abstraction. The interface adopts TDF ports (`sca_tdf::sca_in` and `sca_tdf::sca_out`), usually of type double, to make overall simulation more efficient, and to enforce an efficient interaction with components at any level of abstraction. An example of component implementation is provided in Fig. 7.15, that shows an excerpt of code for the implementation of a battery.

PEUKERT'S LAW

$$C_m = I^n t$$

$$SOC = 1 - \frac{I t}{C_m}$$

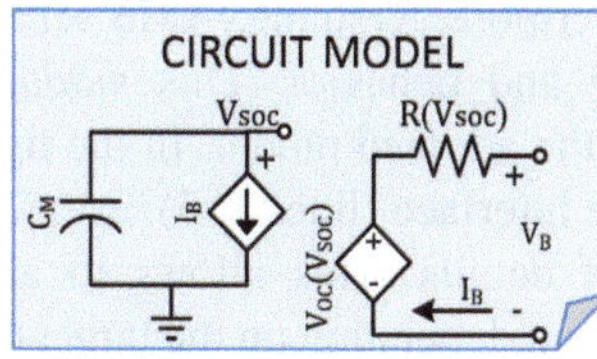


```
SC_MODULE (esd_battery)
{
  sca_tdf::sca_in<double> I;
  sca_tdf::sca_in<bool> En;
  sca_tdf::sca_out<double> V, SOC, E;

  double Vnom, Cm, I0, n;
  ...
};

esd_battery:run(){
  double input, C, soc;

  input = I.read();
  C = Cm * pow((I0/input), n-1);
  soc = 1 - (input/Cm);
  E.write(C);
  SOC.write(soc);
  V.write(Vnom);
}
```

```
SC_MODULE (esd_battery)
{
  sca_tdf::sca_in<double> I;
  sca_tdf::sca_in<bool> En;
  sca_tdf::sca_out<double> V, SOC, E;

  sca_eln::sca_tdf::sca_isource* Ib;
  sca_eln::sca_tdf::sca_vsink* Vsoc;
  sca_eln::sca_c* Cm;
  sca_eln::sca_node n1;
  sca_eln::sca_node_ref gnd;
  ...
  SC_CTOR (battery_lt)
  {
    Ib = new sca_isource("Ib");
    Ib->inp(in);
    Ib->p(n1);
    Ib->n(gnd);
    ...
}};
```

**Fig. 7.15** Example of SystemC-AMS implementation a battery by adopting two different models: Peukert's law [19] (*left*) and a circuit-equivalent model [18] (*right*)

Component evolution is handled differently, depending on the kind of adopted model. Figure 7.15 outlines this idea by comparing two different implementations of a battery. In case of functional models (e.g., waveforms, state machines, or equations), the evolution is handled as a SystemC process, executed at fixed time steps. As an example, the left-hand side of Fig. 7.15 models battery dynamics with Peukert's model [19]. If else the adopted models is at circuit-level [18], it is implemented by describing the circuit as a network of SystemC-AMS ELN components, instantiated and connected in a way that reproduces the circuit specification (right-hand side of Fig. 7.15). Wrapping the ELN subsystem through ELN–TDF converters allows to preserve synchronization with the rest of the system.

Figure 7.15 allows to highlight some of the advantages produced by the adoption of SystemC-AMS as a target language. First of all, a single language allows to cover two very *different levels of abstraction*, i.e., functional and circuit level. Different types of models can thus be simulated simultaneously, leaving the interaction to the underlying simulation kernel. This is advantageous *w.r.t.* co-simulation frameworks.

Furthermore, SystemC-AMS separates the implementation of each component's interface and behavior. This *modularity* allows to preserve the interface when varying the adopted model. In the figure, both implementations of the battery have the same interface (lines 1–5), even if the implemented models are at very different levels of details. This allows to adopt different implementations for the same component, depending on the target (i.e., accuracy or performance), or with the goal of comparing their behavior and characteristics, without affecting the connection to other system components.

Finally, SystemC-AMS modules can be easily *configured and reused* for modeling components with different characteristics. As an example, the SystemC-AMS module can be adopted for modeling two different batteries by setting the different capacity level and all the circuit parameters according to the methodology in Sect. 7.4.3. Thus, the proposed methodology can be enhanced with the definition of a library of models for the components, that can be easily instantiated and configured at later times.

## 7.6 Case Study

This section demonstrates the effectiveness of the proposed framework in the system-level exploration of the power flow of a smart system case study. The framework is indeed compared in terms of simulation accuracy/performance against one of the leading commercial model-based tools, i.e., MATLAB/Simulink [29].

The case study is an industry-strength smart system prototype that consists of integrated devices incorporating sensing, power, computation, and communication into one system. The devices are assumed to be autonomous; this means they have been developed to minimize the power consumption and extend the energy storage lifetimes or to be powered using energy scavenging sources. Figure 7.16 shows the board prototype of the system case study. The building blocks of the system are:

**Fig. 7.16** Board prototype of the smart system case study

- Three load devices including: (1) a set of MEMS sensors such as a 3-axis accelerometer, a MEMS pressure sensor and temperature sensors [25]; (2) a 32-bit ARM-based ST microcontroller used as computing unit to perform data acquisition and control over all the operation of the system [27]; (3) an ST low-power wireless radio for performing wireless data transmission [26]. All these devices are modeled with their execution traces for current and voltage obtained with experimental measurements, as showed in Fig. 7.3.
- A solid state lithium thin film battery by ST, with nominal capacity of 700 μA h and nominal voltage 3.9 V [24]. The battery model is the equivalent circuit depicted in Fig. 7.10.
- A Panasonic stacked coin type supercapacitor with 0.33 F capacitance modeled with the equivalent circuit model showed in Fig. 7.12 [16].
- A DC–DC converter connecting the thin film battery to the power bus, where the conversion efficiency is function of input voltage, output voltage, and current [17].
- Four Texas Instruments TPS63060 DC–DC converters [28] connecting the supercapacitor and the three load devices to the power bus. Their conversion efficiency is function of input and output voltage.
- A PowerFilm Solar MP3-37 flexible PV thin film module made of amorphous silicon cells optimized for indoor applications [20]. The behavior of the PV module was modeled as described in Sect. 7.4.1, using the data from its *current* vs. *voltage* graphs and a daily solar irradiation profile. The model includes also an MPP tracking system in order to generate the maximum power quantity for any irradiance value.
- The power bus modeled as an ideal DC bus consisting of an ideal wire and a voltage generator at the constant reference voltage of 3.0 V. Moreover, the power bus is enhanced with an arbiter that abstracts the power behaviors of the overall system. As long as the power drawn from the PV panel satisfies the power demand of the loads, these are supplied by the power source. Otherwise, the loads are supplied by the supercapacitor during high load power demand intervals, while, during low load power demand intervals, the arbiter activates the thin film battery.

The model architecture of the system case study is shown in Fig. 7.17.

The system model was implemented in SystemC-AMS, by following the implementation process described in Sect. 7.5. Another instance of the system model was implemented in MATLAB/Simulink, using the same models for each device. Simulations have been run by using MATLAB/Simulink R2013a, SystemC 2.3.1, and SystemC-AMS 2.0beta. Both SystemC-AMS and MATLAB/Simulink simulations used a 1 ms timestep, in order to preserve accuracy with respect to the smart system application scenario. Simulations reproduce 10 h of operation of the smart system.

Table 7.2 shows, for both versions of the smart system model, the number of samples (*Samples (#)*) and simulation time (*Time (s)*). Simulation times are calculated as an average over a number of executions. The table shows also the maximum and the average error in the estimation of battery voltage (*Error Max. (%)* and *Avg. (%)*) with respect to MATLAB/Simulink.

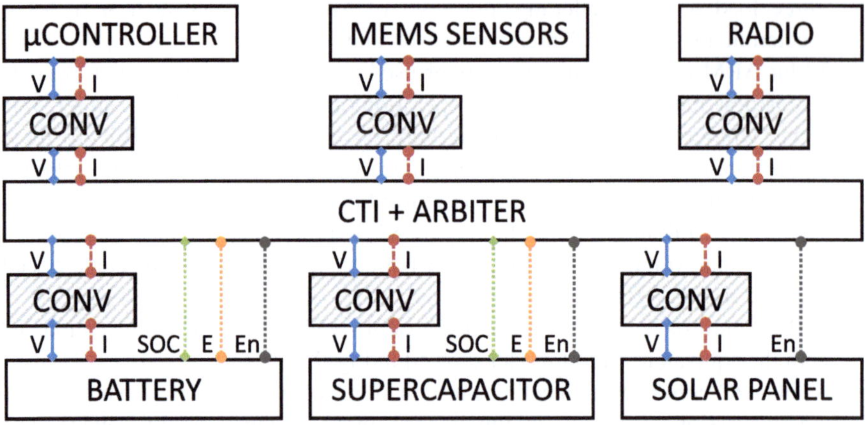


**Fig. 7.17** Model architecture of the smart system case study

**Table 7.2** Accuracy and effectiveness of the proposed methodology w.r.t. MATLAB/Simulink

| | Samples | Time | Error | |
|---|---|---|---|---|
| | (#) | (s) | Max. (%) | Avg. (%) |
| MATLAB SIMULINK | 42,600,002 | 2517.26 | – | – |
| SYSTEMC-AMS | 42,600,020 | 86.80 | 0.04 | 0.0003 |

The SystemC version of the smart system model exhibits very high accuracy with respect to the corresponding Matlab implementation (average error 0.0003 %, maximum error 0.04 %). Concerning speed, SystemC-AMS proved to be 29× faster on average than MATLAB/Simulink (as shown in Table 7.2). Such a slow execution is due to the MATLAB/Simulink internal solver, whose computation is far heavier than the efficient TDF and ELN SystemC abstraction levels.

Figure 7.18 focuses on an excerpt of the overall simulation (50 s) to show the charge allocation policy implemented by the arbiter. In detail, Fig. 7.18b shows the power extracted from the solar panel which is higher on average than the power demanded by the loads, such that part of this power is used to recharge the battery and supercapacitor, as shown by the SOC trends in Fig. 7.18a. In both the scenarios, the waveforms generated by MATLAB/Simulink and by SystemC-AMS coincide, thus highlighting the correctness of SystemC-AMS implementation.

## 7.7 Conclusions

The importance of simulating the flow of power in an autonomous smart device is essential to assess the lifetime of the device or the appropriateness of the sizing of its energy supply sub-system.

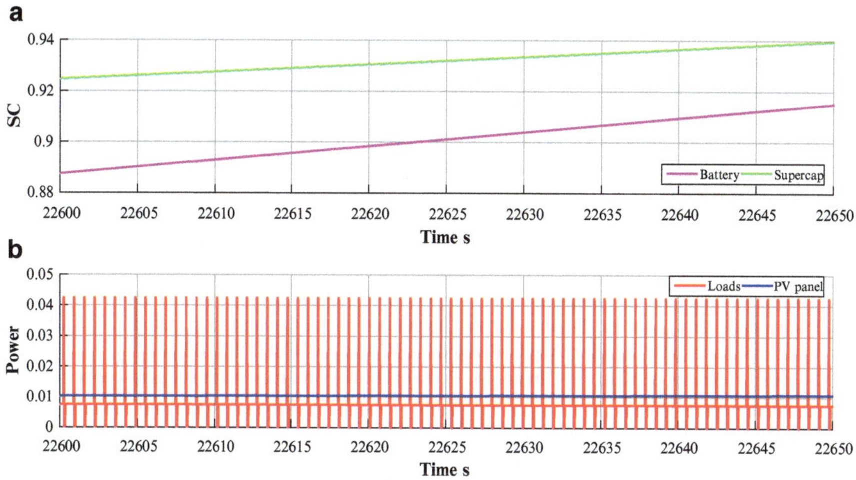


**Fig. 7.18** Excerpt of the simulation of the smart system case study: SOC vs. time for the battery and the supercapacitor (**a**), power vs. time curves for the load devices and the PV panel (**b**)

In this chapter we have proposed an unified approach to the power simulation that, thanks to the use of SystemC extensions to model the non-digital nature of the power flow, allows a designer to concurrently simulate functionality (through traditional SystemC TLM models) of the digital components and the power flow by building and connecting appropriate power models for the "power" components. The advantage of such homogeneous, SystemC-based approach is demonstrated on a real industry-strength device, when a speedup of approximately 30× with respect to Matlab simulations have been observed.

## References

1. Accellera, SystemC (2015), www.systemc.org
2. Accellera, SystemC-AMS (2015), www.systemc-ams.org
3. R. Andoscaa, T. McDonald, et al., Experimental and theoretical studies on MEMS piezoelectric vibrational energy harvesters with mass loading. Sens. Actuators A: Phys. **178**, 76–87 (2012)
4. A. Bauer, J. Hanisch, E. Ahlswede, An effective single solar cell equivalent circuit model for two or more solar cells connected in series. IEEE J. Photovoltaics **4**(1), 340–347 (2014)
5. L. Benini, A. Bogliolo, G.D. Micheli, A survey of design techniques for system-level dynamic power management. IEEE Trans. Very Large Scale Integr. Syst. **8**(3), 299–316 (2000)
6. Canadian Solar, CS6P-250 255 260P Datasheet (2014), www.mitsubishielectric.com
7. M. Chen, G. Rincon-Mora, Accurate electrical battery model capable of predicting runtime and *I*–*V* performance. IEEE Trans. Energy Convers. **21**(2), 504–511 (2006)
8. COMSOL, COMSOL Multiphysics (2015), www.comsol.com
9. Coventor, CoventorWare MEMS Design (2015), www.coventor.com
10. J. Eker, J. Janneck, E. Lee, et al., Taming heterogeneity - the Ptolemy approach. Proc. IEEE **91**(1), 127–144 (2003)

11. P. Fritzson, *Introduction to Modeling and Simulation of Technical and Physical Systems with Modelica* (Wiley-IEEE Press, Hoboken, NJ, 2011)
12. D. Karanjkar, S. Chatterji, et al., An improved current feedback based maximum power point tracking controller for solar photo-voltaic system, in *Proceedings of the IEEE AICERA/ICMiCR* (2013), pp. 1–6
13. W. Lajnef, J.-M. Vinassa, O. Briat, S. Azzopardi, E. Woirgard, Characterization methods and modelling of ultracapacitors for use as peak power sources. J. Power Sources **168**(2), 553–560 (2007)
14. Maxim Integrated, MAX1626 5V/3.3V or Adjustable, 100 % Duty Cycle, High-Efficiency, Step-Down DC–DC Controllers - Maxim (2015), www.maximintegrated.com/en/products/power/switching-regulators/MAX1626.html
15. MIDE, Volture EHE004 Datasheet (2013), www.mide.com
16. Panasonic Corp., Electric Double Layer Capacitors (Gold Capacitor) Catalog (2015), http://industrial.panasonic.com/lecs/www-data/pdf/ABC0000/ABC0000PE102.pdf
17. S. Park, Y. Wang, Y. Kim, N. Chang, M. Pedram. Battery management for grid-connected PV systems with a battery, in *Proceedings of the 2012 ACM/IEEE International Symposium on Low Power Electronics and Design* (2012), pp. 115–120
18. M. Petricca, D. Shin, et al., An automated framework for generating variable-accuracy battery models from datasheet information, in *Proceedings of the ACM/IEEE ISLPED* (2013), pp. 365–370
19. W. Peukert, Über die Abhängigkeit der Kapazität von der Entladestromstärke bei Bleiakkumulatoren, in *Centralblatt fur Elektrotechnik* (1897), p. 20
20. PowerFilm, MP3-37 Datasheet (2015), www.flexsolarcells.com/PowerFilm-Solar-OEM-Components.php
21. R. Rao, S. Vrudhula, D. Rakhmatov, Battery modeling for energy aware system design. IEEE Comput. **36**(12), 77–87 (2003)
22. M. Singh, S. Santoso, Dynamic Models for Wind Turbines and Wind Power Plants. Technical report, National Renewable Energy Laboratory (May 2011) (2015), www.osti.gov/bridge
23. SpectroLab, GaAs/Ge Single Junction Solar Cells Datasheet (2015), www.spectrolab.com
24. STMicroelectronics, ST EFL700A39 EnFilm rechargeable solid state lithium thin film battery datasheet (November 2013)
25. STMicroelectronics, MEMS and Sensors (2015), http://www.st.com/web/en/catalog/sense_power/FM89?icmp=fm89_pron_memsx2_sep2013
26. STMicroelectronics, SPIRIT1 Low Data Rate, Low Power Sub-1GHZ Transceiver (2015), http://www.st.com/web/en/resource/technical/document/datasheet/DM00047607.pdf
27. STMicroelectronics, STM32 32-bit ARM Cortex MCUs (2015), http://www.st.com/web/en/catalog/mmc/FM141/SC1169?sc=stm32
28. Texas Instruments, Inc., Texas Instruments TPS63060 DC–DC Converter (2014), http://www.ti.com/product/tps63060
29. The Mathworks Inc., Simulink – Simulation and Model-Based Design (2015), http://www.mathworks.com/products/simulink
30. C. Unterrieder, M. Huemer, et al., SystemC-AMS-based design of a battery model for single and multi cell applications, in *Proceedings of the IEEE PRIME* (2012), pp. 1–4
31. S. Vinco, A. Sassone, et al., An open-source framework for formal specification and simulation of electrical energy systems, in *Proceedings of the IEEE/ACM ISLPED* (2014), pp. 287–290
32. M.S.M. Wens. Basic DC–DC converter theory, in *Design and Implementation of Fully-Integrated Inductive DC–DC Converters in Standard CMOS* (Springer, New York, 2011), pp. 27–63
33. L. Zubieta, R. Bonert, Characterization of double-layer capacitors for power electronics applications. IEEE Trans. Ind. Appl. **36**(1), 199–205 (2000)

# Chapter 8
# Smart System Case Studies

**Ignazio Blanco, Fabio Cenni, Roberto Carminati, Angelo Ciccazzo, Sandro Dalle Feste, Franco Fummi, Giuliana Gangemi, Fabio Grilli, Michelangelo Grosso, Mirko Guarnera, Michele Lora, Anna A. Pomarico, Gaetano Rasconà, Salvatore Rinaudo, and Giuditta Roselli**

## 8.1 Introduction

The design of today's smart systems requires integration-aware, component-level design, modeling, and optimization methodologies applicable to different domains, i.e., analog and digital electronics, MEMS, discrete and power devices and software.

I. Blanco • A. Ciccazzo • G. Gangemi • M. Guarnera • S. Rinaudo
STMicroelectronics s.r.l., Catania, Italy
e-mail: ignazio.blanco@st.com; angelo.ciccazzo@st.com; giuliana.gangemi@st.com; mirko.guarnera@st.com; salvatore.rinaudo@st.com

F. Cenni
STMicroelectronics SA, Crolles, France
e-mail: fabio.cenni@st.com

R. Carminati • S. Dalle Feste • F. Grilli
STMicroelectronics s.r.l., Agrate Brianza (MB), Italy
e-mail: roberto.carminati@st.com; sandro.dallefeste@st.com; fabio.grilli@st.com

F. Fummi
EDALAB s.r.l., Verona, Italy
e-mail: franco.fummi@edalab.it

M. Grosso (✉)
ST-Polito s.c.ar.l., Torino, Italy
e-mail: michelangelo.grosso@st-polito.com

M. Lora
Università di Verona, Verona, Italy
e-mail: michele.lora@univr.it

A.A. Pomarico • G. Roselli
STMicroelectronics s.r.l., Lecce, Italy
e-mail: anna.pomarico@st.com; giuditta.roselli@st.com

G. Rasconà
ST-Polito s.c.ar.l., Catania, Italy
e-mail: gaetano.rascona@st-polito.com



N. Bombieri et al. (eds.), *Smart Systems Integration and Simulation*,
DOI 10.1007/978-3-319-27392-1_8

As a matter of fact, in each design domain there exist industrial design and implementation flows leading from the specifications to a product; however, in the innovative and increasingly feature-filled devices required by the market, largely different components have to be assembled and squeezed into extremely small volumes, and work together to attain the desired system behavior.

Miniaturized heterogeneous systems require the evaluation of functional and non-functional, multiphysical interactions between the components and with the environment: therefore, the integration of component and sub-system models in a comprehensive simulation environment enabling fast performance evaluation and design space exploration is more and more desirable. The effectiveness and efficiency of the component-level design and modeling techniques have been demonstrated and validated in the various test cases reported in the other chapters of this book. This chapter aims at illustrating the system-level integration issues introducing two case studies that integrate heterogeneous components: a laser beam steering (LBS) pico-projector actuator and a wearable sensor node for reliable drift-free limb tracking. In both cases, performance optimization is dependent on the accurate and manageable representation of the different interactions between hardware and software parts in mission-like conditions, enabled by model abstraction and co-simulation methodologies, and that in this case is provided by the SMAC Platform [1–3]. In addition, this chapter presents the description of an archetypal smart system, an *open-source test case* (OSTC), on which model creation and abstraction are demonstrated, and low- and high-level simulation methodologies are largely employed. The OSTC has also a tutorial value, as its models are freely downloadable and distributable to demonstrate the abstraction and simulation tools abilities.

The chosen case studies potentially cover several key aspects relevant for the co-simulation of multi-layer/multi-scale and multi-domain physical interactions:

- Vibrations/movement. The horizontal and vertical deflecting mirrors within the pico-projector actuator are driven at different frequencies (up to resonance) and their movement depends on physical features and control algorithms. The wearable sensor node has to acquire physical parameters and track limb position from static to relatively low-frequency dynamic conditions.
- Fluid dynamics. In the pico-projector actuator, the air molecules around the mirrors have to be considered as a fluid.
- Acoustics. At a lower scale, the effect of air drag (the torsion of the fingers of the micromirror in the pico-projector actuator) due to the air heating has to be studied as an acoustic phenomenon.
- Temperature. The high-frequency movement and the employed laser source cause heating in the pico-projector actuator, so the movement in the hot air surrounding the actuating devices has to be studied as a thermo-mechanical issue. In addition, in specific implementations, the sensors are integrated into a large assembly, which can self-heat. Therefore, the nanodevices, as well as the full system, have to self-tune with temperature to keep the measurement accuracy

well controlled. Alternatively, temperature can be measured in the same package to monitor the sensor operating conditions.

- Magnetic interactions. In the development of the wearable sensor node it is needed to account for "hard iron" and "soft iron" effects.

For each case study, this chapter presents a general system description and related requirements, the chosen model implementation strategy, and the achieved results. In addition, the OSTC section provides a tutorial on the use of model abstraction tools for smart system design and simulation.

## 8.2 Virtual Prototyping Platform for a Pico-Projector Actuator

### *8.2.1 General System Description and Requirements*

Pico-projectors are miniature video projectors that can be used in many applications where there is a need for sharing information over large displays [4]. They provide a convenient way to project quickly any kind of optical information onto any surface such as walls, notebooks, or airplane food trays, as well as a new array of portable projection screens sized and designed for pico-projectors and presenter mobility. Such a technology is based on systemic miniaturization and integration of heterogeneous technologies, functions, and materials. This characteristic makes pico-projectors a suitable target for the implementation of smaller, smarter (predictive, reactive, and cognitive), and energy-autonomous smart systems.

In this case study, developed by STMicroelectronics, laser is used as a light source, instead of currently more frequent Digital Light Processing (DLP) or Liquid Crystal on Silicon (LCoS) technologies for optical systems. In fact, laser-based projectors have several advantages compared to the other technologies, such as a smaller device size, brighter colors, higher efficiency, and the possibility of having a continuously focused picture and to project at variable distances and onto curved surfaces, without the need to keep turning a focus dial.

Such pico-projectors are based on laser sources of the three primary colors (red, green, and blue) that are modulated and their light converges into a single beam. This beam is then deflected by a mirror system thus producing a light spot that is scanned and projected across the image plane, similarly as the electron beam in a cathode ray tube. A laser pico-projector is an extremely complex system, whose main components are micromirrors, high voltage mirror drivers, laser drivers, and a video digital processor and mirror controller.

The case study exploits state-of-the-art technologies, for both the mechanical micromachining and the ASIC control circuitry. Some of the key technological issues involved in the pico-projector development include:

- Micromirrors suitable for pico-projection applications, which are fabricated exploiting the emerging technology of MOEMS (Micro-Opto-Electro-Mechanical Systems) for accurate angular deflection of light beams at millimeter and submillimeter scale.
- Mirror drivers design, requiring high voltage and implemented by means of the BCD8s SOI 190V technology (channel: 0.35 μm; Logic supply: 3.3 V; Max high voltage supply: 190 V).
- High performance laser drivers, employing BCD8st technology.

In addition to this, the pico-projector project requires different kinds of expertise to support modeling, investigation and validation using a variety of languages and simulation instruments: from mechanical description tools, to analog and digital design software, to top-level integration languages.

The design activities involved the development of the single components, to obtain the expected behavior from the resulting devices, and then the construction of a comprehensive system-level model, fundamental to assess the system-level performance and, more specifically, to evaluate and compensate the *ripple* effect. Ripple is a real-world problem affecting laser pico-projectors, due to the beam steering mechanisms: When electromechanical mirrors are used, the horizontal deflection is performed by horizontally rotating the reflecting surface at its natural frequency. Conversely, vertical scanning is obtained by vertically rotating the mirror with a linear voltage at a lower speed. Each time a complete frame is scanned, the retrace of the mirror to the initial position induces vertical vibrating modes leading to non-linear effects in the trace time. The visible effect is an uneven distribution of the distance between the traced horizontal lines, causing a number of lighter and darker horizontal stripes on the projected image. A carefully developed compensation algorithm can be used on the controller to eliminate or, at least, to mitigate this effect. Such algorithm was developed on the complete system virtual model.

### 8.2.2 Model Implementation

MEMS micromirrors have been designed and carefully characterized by means of mechanical simulations, using a variety of software tools such as ANSYS, COMSOL Multiphysics, and Coventor MEMS+. These tools have been widely exploited in order to optimize technological process flow and mirror design, so as to obtain the required actuation force and resonance frequencies.

VHDL-AMS/Verilog-AMS and MATLAB Simulink models for the MEMS components, the power electronics, the analog and conversion circuitry (transimpedance amplifier—TIA, analog-to-digital ADC, and digital-to-analog DAC converters), and the digital control logic (Digiboard) were developed starting from different simulator and data sources. Then, these descriptions were translated to SystemC/SystemC-AMS to obtain a homogeneous system-level model, using the

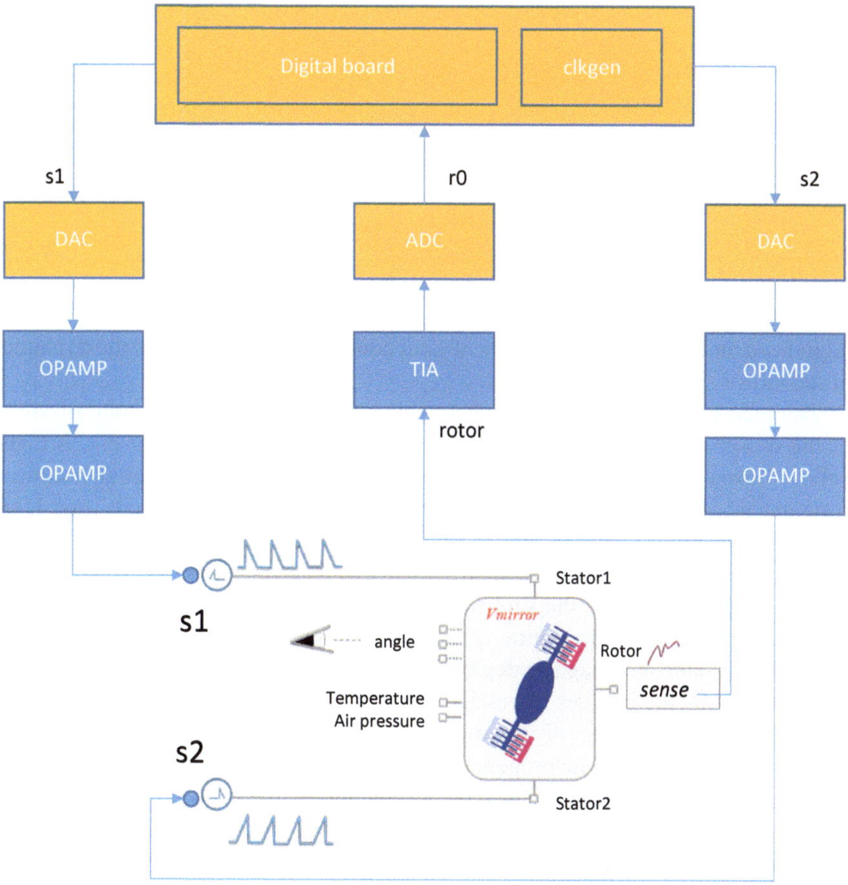


Fig. 8.1 Vertical scanning mirror and related driving and control circuitry

SMAC Platform abstraction tools. Figure 8.1 presents a comprehensive view of the system targeted by the model.

System-level simulation, based on a single simulation engine, can be efficiently performed on this description. In order to evaluate and minimize the ripple effect with the use of a prototype, the evaluation of the ripple magnitude would require either a local luminance measuring system or the tracking of the mirror rotation angle, however, the latter cannot be monitored unless a complex video sensing-based processing is implemented. The developed model, instead, allows implementing and evaluating the control algorithm executed by the digital control logic (using either open-loop or closed-loop approaches), which can be modified and evaluated considering the deflection angle $\theta$ and its velocity variations (Fig. 8.2). In order to measure and compensate the ripple effect, the current flowing through the mirror rotor is used as indicator. The current is amplified by means of the TIA and

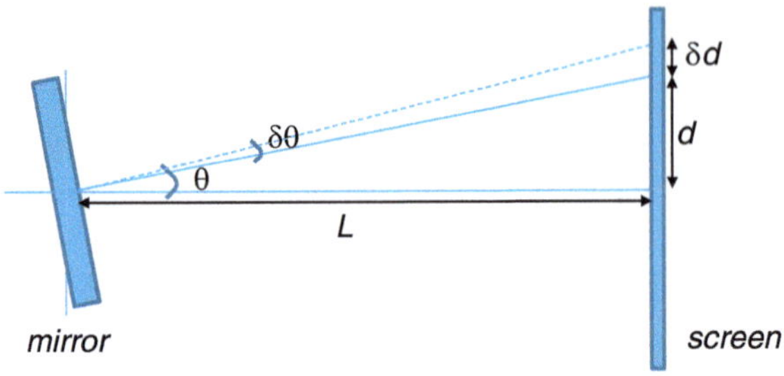


**Fig. 8.2** Schematic representation of the Vertical Mirror deflection and its effect on projection: ripple is minimized with constant $\delta d/\delta t$, in turn related to the speed of the deflection angle $\theta$

digitalized and then processed by the digital control logic that updates the dynamics of the mirror stators driving signals.

The pico-projector is a highly heterogeneous system, from both point of views of the design domains and the physical laws involved. Referring to Fig. 8.1, the yellow boxes (i.e., Digiboard, clkgen, DACs, and ADCs) are digital hardware components, synchronous to a digital clock signal. Moreover, the ADC and DAC are also presenting electrical-linear statements for digital to analog and analog-to-digital conversion. The blue boxes (i.e., OPAMP and TIA) represent purely analog components, while the Vmirror component is the electro-optical sub-system. Signals *s1*, *s2*, and *r0* are digital quantities characterized by a discretized value. Meanwhile, all the other signals depicted in the figure are electrical interconnections characterized by their values of voltage and current flow.

Furthermore, the type of equations implemented by the single component gives another source of heterogeneity among different sub-parts of the system. Figure 8.3 classifies the different types of equations that can be used to describe a time-invariant physical system. Then, it suggests a possible mapping of such equations into the SystemC-AMS models of computation (MoCs). The red arrows of Fig. 8.3 represent the most direct theoretical mapping. It consists in mapping electrical-linear equations in the Electrical Linear Network (ELN) MoC, anything non-linear into the Timed Data Flow (TDF) MoC, and using linear signal flow (LSF) to represent equations that are linear but not electrical. The green arrows represent a performance oriented mapping. The TDF MoC is more performant with respect to the others since it exploits a static scheduler, faster than the dynamic one used by ELN and LSF. Thus, the performance-oriented mapping depicted by the green arrows is trying to maximize the use of this MoC instead of LSF and ELN.

The sub-components of the system have been characterized according to the taxonomy in Fig. 8.3, and their classification is depicted in Fig. 8.4. Then, according to the taxonomy, every single component has been mapped into the most suitable MoC. Furthermore, since one of the targets of the SystemC-AMS model is to speed-up the simulation with respect to the variety of tools used to create the initial model, the performance-oriented mapping has been applied whenever possible. Then the translation to SystemC-AMS has been done starting from the available Verilog-AMS and MATLAB Simulink models of each single component of the pico-projector, and each model was thoroughly validated.

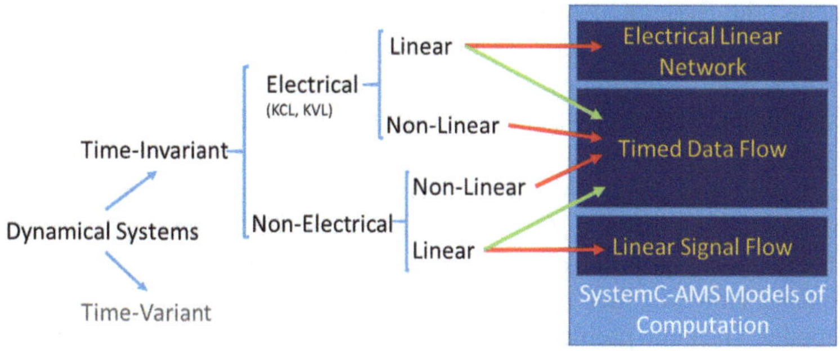


**Fig. 8.3** Systems classification and mapping in SystemC-AMS models of computation

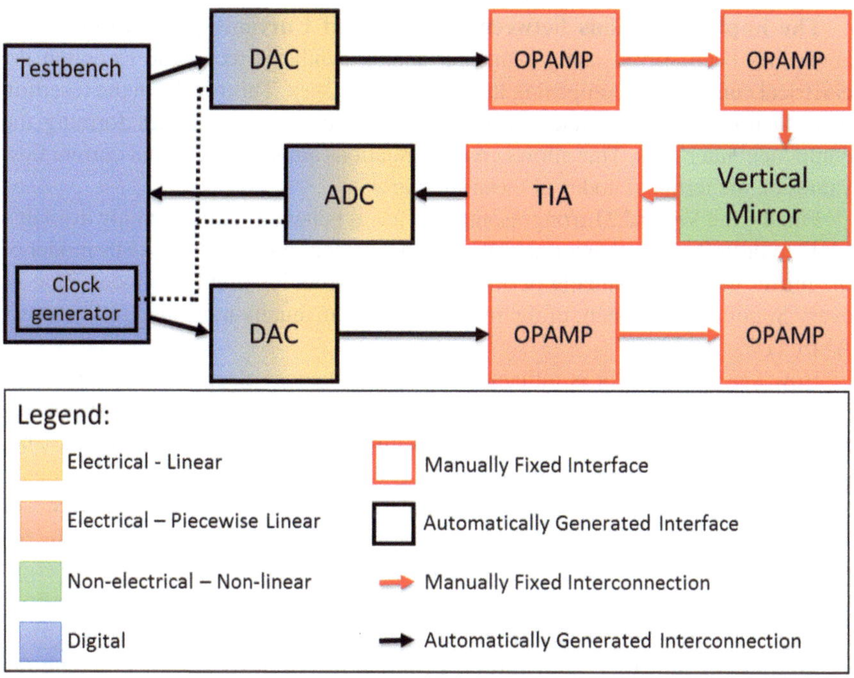


**Fig. 8.4** Classification of the components of the pico-projector, according to the implemented equations

The translation of the digital components (i.e., Digital Board and Clock Generator) relies on HIFSuite front-end and back-end tools and it has been automatically performed. In particular, the components are translated in HIF format using *verilog2hif* and converted in SystemC by *hif2sc*.

Converters (i.e., ADC and DAC) present only linear-electrical equations and digital RTL statements. The digital sub-parts have been translated into SystemC as for the purely digital components described above. Concerning the electrical statements, the SystemC-AMS description is based on the instantiation of basic components. Every instantiated component contributes with a set of equations to the system. The methodology aims at reproducing the original set of equations expressed in the Verilog-AMS model connecting the instantiated components. The methodology is implemented on top of HIFSuite.

The set of components presenting electrical-piecewise linear equations that are the operational amplifier (i.e., OPAMP) and the transimpedance amplifier (i.e., TIA) have been initially translated using the automatic tool used for electrical-linear components. However, their composition was not equivalent to the original Verilog-AMS model. The main problem is due to the *implicit relations between Voltages and Currents* that are often present on the interface terminal nodes of the components. These relations are not considered by the methodology previously presented. Thus, some manual modifications have been performed.

**The implicit relations between Voltages and Currents problem** described above has been solved using a manual analog synthesis technique to define the electrical components composing the module interfaces. This results in the insertion of capacitors, inductors, and resistances nearby the terminal nodes forming the component interfaces. This allows creating relations among voltage and current flow values on the terminal nodes of these sub-parts.

Finally, the Vertical Mirror presents non-linear behavior. Thus, the only available MoC to describe its behavior is TDF. For this reason, the SystemC-AMS model of the mirror has been manually performed starting from its Simulink model. That is, every Simulink component of the model has been manually translated in SystemC-AMS TDF.

However, its behavior is influenced by and influences electrical relations on its input/output ports and on the input/output terminal of the surrounding components (i.e., OPAMP and TIA), thus introducing bi-directional relations. On the other hand, since TDF belongs to the dataflow family of computational models, it allows only specifying unidirectional information flows. Thus, modeling the system in terms of TDF primitives leads to a *loss of interface information* about the relations among the physical quantities involved in the system, i.e., the incapability to express an energy-conservative behavior hence to keep a relation between effort and flow quantities (voltage and current in the electrical domain).

**The loss of interface information problem** has been solved by wrapping the TDF model within an interface described in terms of ELN MoC. Furthermore, the sub-part of the TDF model corresponding to the capacitors used in the interface has been replaced with the instantiation of controlled capacitors of the ELN MoC, within the ELN wrapper. Then, the model has been connected to the other parts of the system. In such way, it was possible preserve the relations lost during the conversion from the electrical Simulink description to the TDF one.

In order to improve the performance of the system simulation, also for linear and piecewise linear components a set of manual manipulations have been performed in order to maximize the use of TDF.

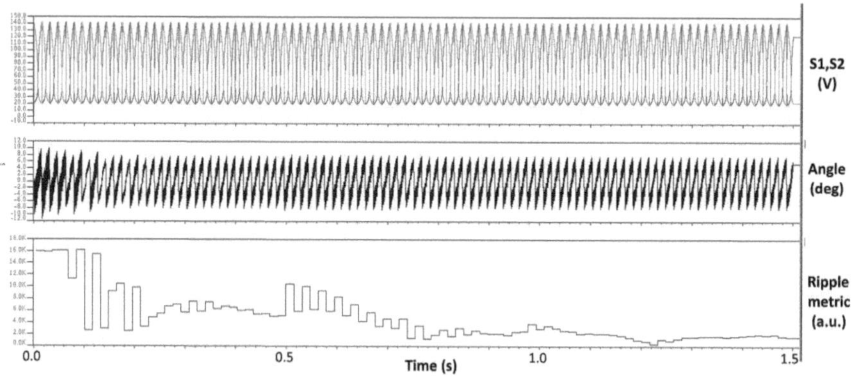


**Fig. 8.5** Closed-loop top-level waveforms with ripple compensation on the vertical axis

In particular, for both the OPAMP and the TIA, a sub-set of the internal equations has been manually resolved and re-written using SystemC-AMS TDF. This rewriting is based on the identification of chains of assignments reproducing the relations between sub-sets of nodes of the single component. Of course, for the reasons introduced before, these sub-sets of nodes cannot comprehend any terminal node used in the interface of the components.

In conclusion, using the set of tools provided within HIFSuite it was possible to automatically translate the digital RTL descriptions, the electrical-linear descriptions and the internal parts of the electrical-piecewise linear components. While, manual support was necessary for the non-electrical. Figure 8.4 gives also an overview of the state of the translation automation. Parts filled in purple and yellow have been automatically translated. The Vertical Mirror (green box) has been translated in a completely manual way. Components depicted by orange boxes have been partially automatically translated. The interfaces of the components represented by black margin boxes have been automatically generated without any manual intervention, while red margins represent components whose interfaces have been manually fixed after the automatic or manual conversion. Finally, black arrows represent automatically generated interconnections between components, while red arrows represent interconnections that required manual interventions to be fixed.

### 8.2.3 Experimental Results

Figure 8.5 shows the output of a top-level simulation of the overall platform and the compensation of the ripple using the developed algorithms, following the approach outlined in [5]: from top to bottom, the driving voltages (S1 and S2), the deflection angle and the ripple magnitude (in arbitrary units) are shown. The last value is

**Table 8.1** Overall simulation time

| Modeling style | Pico-projector virtual platform (open loop) | Pico-projector virtual platform (closed loop) |
|---|---|---|
| Verilog/VerilogA/Verilog-AMS (Questa-AMDS) | 5 s (tuning = STD or FAST)/30 ms (166) (no tracing) | NA |
| SystemC/SystemC-AMS | 4 s (tracing all signals)/30 ms (133)<br>1.5 s (no signal traced)/30 ms (50) | 177 s (top-level tracing)/1.5 s (118)<br>74 s (no signal traced)/1.5 s (49) |

computed as a function of the current measured by the TIA, by evaluating the characteristics of the signal on a number of consecutive samples. The simulation lasts 1.5 s and it shows that the ripple is reduced considerably during the simulation time thanks to the iterative optimization applied by the developed algorithm.

The overall simulation times are given in Table 8.1, with the following convention: <simulation time>/<simulated time> (simulation/simulated time ratio).

The Verilog-based model is slightly slower than SystemC-based implementation since it is constrained by the analog solver time step (eldo for Questa-ADMS, a tuning can be done) but the Verilog-based approach does not allow to simulate the C/C++ embedded SW in the loop. The developed Simulink model does not allow the modeling/simulation of the entire system.

The SystemC-AMS approach does not require license-fees since the simulator is distributed under Apache 2 license by the Fraunhofer Institute.

In conclusion, the virtual prototyping platform for the pico-projector actuator proved to be useful for the performance evaluation of the system and for the optimization of the control algorithms driving it. Due to the complexity and the heterogeneity of the device, no suitable models were previously available to perform a complete simulation, including the electromechanical, analog/power, and digital control parts, and taking into account the effects of the non-ideality of each part. Traditional approaches for evaluating projection quality metrics and for optimizing the system performance required the use of expensive prototypes and laboratory setups, exploiting time-consuming and sub-optimal trial-and-error methodologies. The virtual prototype developed using the model abstraction approach provided a solution for mitigating the ripple problem and a viable approach for the development of heterogeneous integrated systems involving multiphysical aspects. It has to be underlined that the adopted methodology is based on the abstraction of lower-level, detailed models of the system components and on a top-level *simulation* involving a single engine and thus optimizing computation-communication cost.

## 8.3 Wearable Sensing Equipment for Reliable Drift-Free Limb Tracking

### *8.3.1 General System Description and Requirements*

The monitoring of real limbs trajectories in daily life experiences is attractive for many application fields: fitness, sport, rehabilitation, and healthcare and wellness in general. Market available solutions for motion capture are mostly divided into optical trackers and inertial measurement units (IMUs). The recent advances in inertial sensors allow a high level of integration of an IMU solution with very low cost and quite good measure reliability. In addition, compared to optical methodologies, they do not require setting up a specific infrastructure in the measurement environment and do not lose motion tracking when a body part is hidden from the camera.

The proposed system is composed of a set of inertial sensor nodes fixed to different body parts (e.g., arm and forearm), whose angular positions are determined and transmitted to a central unit able to reconstruct the positioning and relative movement of limbs. An important part of the data elaboration (the so-called *sensor fusion*, providing attitude and orientation angles) is performed on each node, thus constituting a distributed system architecture.

The system requirements are the following:

- Use of sensor nodes, each one equipped with tri-axial gyroscope, tri-axial accelerometer, and tri-axial magnetometer.
- Each sensor node has to perform a preliminary sensor fusion based on extended Kalman filtering (EKF).
- Each sensor node has to compensate temperature effects (between 0 and 50 °C) and magnetic disturbance caused by vicinity to ferromagnetic elements.
- The sensors shall transmit real-time data to a central unit able to display their relative positions in space.
- The maximum current consumption of each sensor node shall be less than 100 mA at 3.3 V (powered by battery).
- The tracking accuracy (computed limb angle in static condition) shall be $< 2°$.

The developed system (Fig. 8.6a) is made of sensor nodes based on the iNEMO M1 system-on-board (SoB), shown in Fig. 8.6b, which includes three types of inertial sensors:

- A tri-axial MEMS gyroscopic sensor, providing real-time angular velocity readings (ST L3GD20);
- A tri-axial MEMS accelerometer, able to measure the physical acceleration associated with the phenomenon of weight experienced by any test mass at rest in the frame of reference of the accelerometer device, and therefore summing up gravity and motion effects;
- A tri-axial MEMS magnetic field sensor.

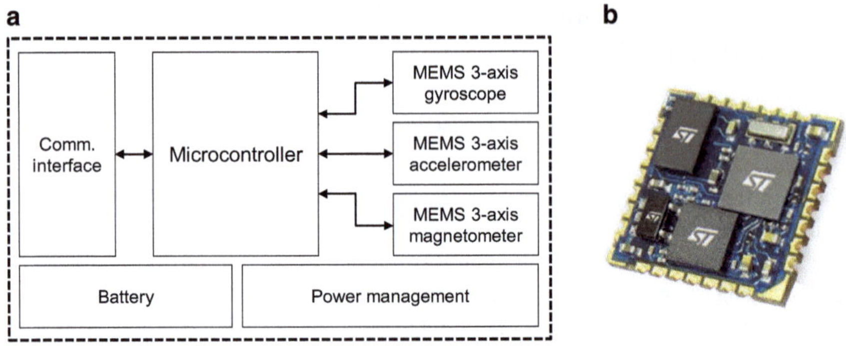


**Fig. 8.6** Block diagram of a sensor node (**a**) and the iNEMOM1 system-on-board (**b**)

The latter two devices are encapsulated in a single package (ST LSM303DLHC). Additionally, the SoB includes a 32-bit microcontroller (STM32F103) able to preliminary filter and elaborate data coming from sensors to perform a first-level sensor fusion. In static condition, the accelerometer is useful for determining the vertical direction associated to the gravitational acceleration, while the magnetometer provides an indication of orientation with respect to the magnetic North. Gyroscope readings are integrated to compute angular displacements, which are typically affected by drift that is periodically corrected thanks to the output of the other sensors.

For the development of a reliable and drift-free system, a large number of details and the device non-ideality need to be taken into account in the fine-tuning of the involved algorithms. This includes, for instance, "hard" and "soft iron" calibration for the magnetometer, which should be insensitive of local perturbations of the Earth's magnetic field, and the compensation of the effects of temperature on the sensor transfer functions.

A precise yet agile high-level system model capturing the behavior of the selected components is of paramount importance to optimize the system functionalities before prototyping, in mission-like conditions. As a matter of fact, the usual laboratory setup for performance characterization of such a system includes costly vision-based equipment employing a number of cameras, markers on the body of a person, and complex motion tracking algorithms (e.g., [6]). With respect to the state-of-the-art, the new abstraction-based methods and tools enable significant cost reduction in application development and the study and evaluation of new product ideas.

To this purpose, the current approach is based on the creation of a functional model of a sensor node in the Keysight SystemVue framework, composed of the sensor models in SystemC/SystemC AMS and the sensor fusion algorithm code that will be run by the microcontroller. The proposed approach includes also the use of a limb movement simulator environment for synthesizing suitable inputs for the sensor models while injecting possible environmental perturbations. The high-level sensor models can be developed resorting to automated abstraction techniques, or directly written in one of the standard formats supported by the platform.

### 8.3.2 Model Implementation

The objective is the development of a functional model of the iNEMO M1 SoB, including the MEMS sensors and the algorithms running on the embedded microcontroller [7]. This model is needed to evaluate the performance of sensor fusion algorithms and to investigate the improvements that can be applied. For this reason, digital communications at the system level are handled as transactions, and instead of modeling the complete microcontroller, the same C language sensor fusion library that will run on the real system is instantiated as a code block in the simulation framework.

The model was assembled in the SystemVue environment and includes the following blocks:

1. SystemC/SystemC AMS model of the LSM303DLHC tri-axial accelerometer and tri-axial magnetometer.
2. SystemC/SystemC AMS model of the L3GD20 tri-axial gyroscope.
3. Sensor fusion algorithms and system management functions.

All the MEMS sensor devices are characterized by the integration of the micromechanical sensing part, an analog front-end taking care of power supply, self-regulation, amplification, filtering and analog-to-digital conversion, and a digital front-end (DFE). Thus, they can be considered sub-systems including MEMS, analog and digital parts. Figure 8.7 shows the block diagram of the employed accelerometer model for one axis. Its main composing elements are the mechanical second order Laplace transfer function, expressing the relationship between applied acceleration (and consequently force *F*) and the displacement *d* of the oscillating mass, a non-linear module expressing the capacitance of the plates *C* as a function of their displacement, the analog front-end (AFE, comprising amplifiers, and filters), the analog-to-digital converter (ADC), and the DFE. As a matter of fact, mechanical, analog, and digital electronic domains are represented [8].

The model parameters derive from lower-level representations employed by the device designers: they represent the behavior of a "nominal" device, but they can

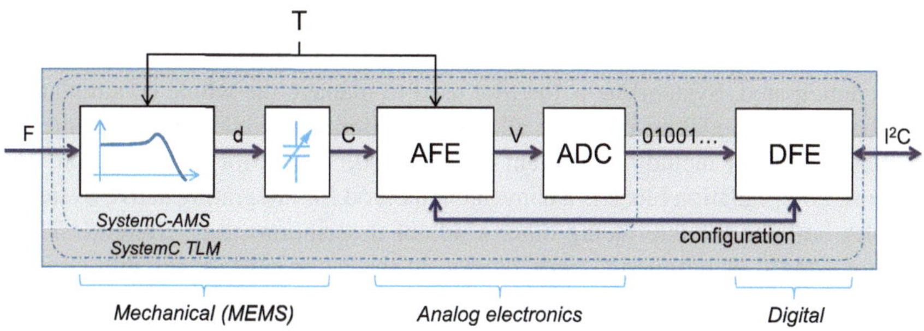


**Fig. 8.7** Accelerometer model (for one axis)

be easily customized in order to consider the effect of manufacturing variability and corner cases or to mimic specific devices. Beyond the functional multi-physics relations that allow sensing acceleration and expressing it as a digital reading, noise is also simulated. The temperature value is taken into account in a limited range (0–50 °C) and its changes dynamically affect the sensor model behavior. As it appears from the figure, the mechanical and analog parts, up to the ADC module, are described in SystemC-AMS language using the TDF model of computation. SystemC TLM is used to describe the behavior of the DFE including the state machines that manage the device read-out configuration and the programmable registers (full scale selection, output data rate, and digital filters). The magnetometer (sharing its digital interface with the accelerometer within the LSM303DLHC module) and the gyroscope are to be modeled in similar ways.

Regarding the sensor fusion algorithm, the iNEMO Engine Lite library developed and distributed by STMicroelectronics [9] has been used to define an initial setup. It applies an EKF algorithm. In general, the EKF algorithm addresses the problem of estimating the state of a discrete-time process described by the equations below [10]:

$$x_{k+1} = A_k x_k + B_k u_k + w_k$$
$$z_k = H_k x_k + v_k$$

where $k$ is the step time, $x$ is the state vector, $u$ is the input vector, $A$, $B$, and $H$ are the state, input, and output matrixes, respectively, and $w$ and $v$ are state and measurement noise. The latter are represented by Gaussian white noise sources with covariance matrixes $Q$ and $R$, respectively. The flow diagram for the recursive algorithm is shown in Fig. 8.8.

At each time step, the algorithm propagates both a state estimation $x_k$ and estimation for the error covariance matrix $P_k$. The latter provides an indication of the uncertainty associated with the current state estimation. These are evaluated in the predictor equations. The Kalman gain $K$ is derived by minimizing the a posteriori error covariance and could be considered a measurement of the confidence level of the predicted state.

Figure 8.9 presents the block diagram of the system-level simulation model for a sensor node. As it appears from the diagram, multi-physics effects are taken into account in the system behavior simulation in terms of sensor functional inputs (acceleration, magnetic field, and angular velocity) and temperature.

As anticipated, SystemVue is the tool used to manage the whole co-simulation. Its block-based modeling interface allows instantiating modules from the internal software library or including different kinds of code blocks for co-simulation. The SystemC co-simulation block is a convenient method for integrating native SystemC modules but also for interfacing other kinds of descriptions, e.g., SystemC TLM, SystemC AMS, or untimed C code blocks. In the latter cases, a suitable wrapper has to be developed for providing the needed signal conversions and defining block timing. The sensor fusion algorithms are implemented as a multiplatform C library that can run on PC as well as on the ARM core of the STM32 microcontroller family. The functions are to be instantiated within a SystemC co-simulation block

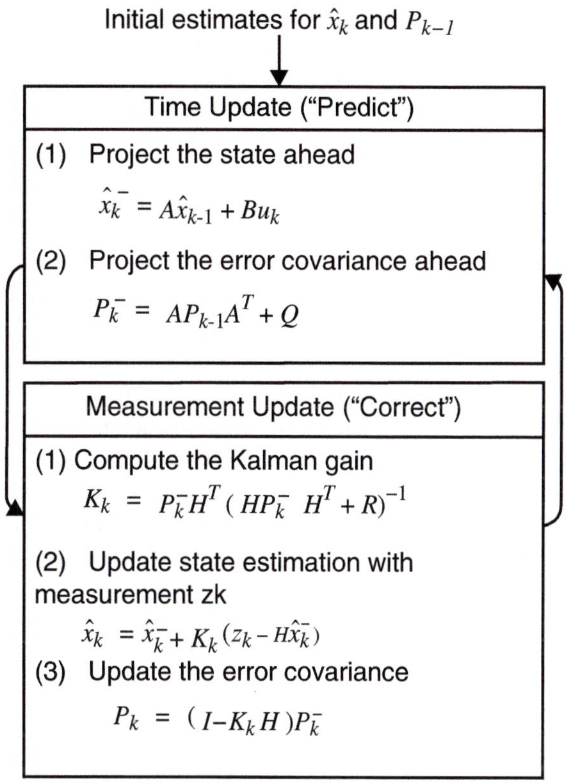


**Fig. 8.8** Flow diagram of the time-discrete EKF algorithm [11]

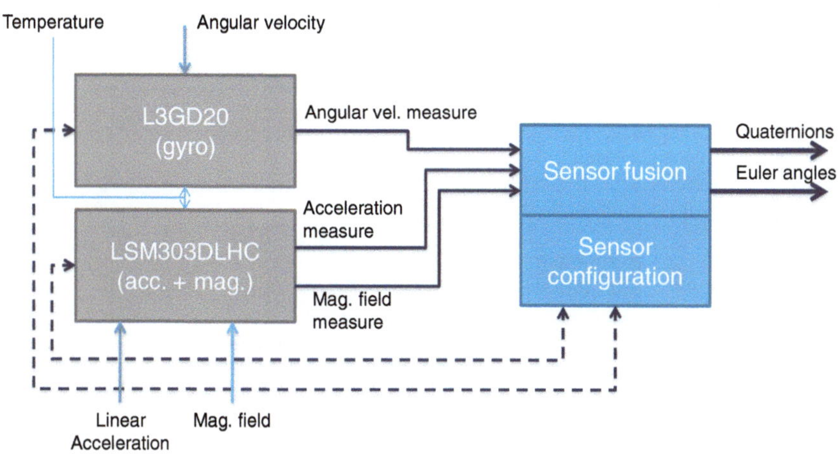


**Fig. 8.9** System-level simulation model for a sensor node

that implements the required state update (quaternions and therefore Euler angles) at each sensor reading cycle. The algorithm can be modified directly in the simulation framework, in order to verify the effects of changes on the complete system. Additional functions can be added to work either on the raw data coming from the sensors for filtering noise and compensating environmental disturbs, or on already elaborated results.

Figure 8.10 shows the system-level simulation and analysis setup for a sensor node. OpenSim, an open-source platform for modeling, simulating, and analyzing the neuromusculoskeletal system [12], was used for generating input stimuli for the model. The tool allows extracting kinematic data about the location in which each virtual sensor is placed, either with real medical data (available, e.g., in the SimTK databases from Stanford University) or with simulated motions. The BodyKinematics and PointKinematics analysis tools are used to get the orientation, the angular speed, and the position of each body part hosting a sensor (e.g., tibia, femur, pelvis) while the simulated body is moving. The orientation of the body part (i.e., of the node, in the node's reference system) is used to suitably decompose on the node rotated axes the values of gravity, linear acceleration, and magnetic field in order to properly feed the accelerometer and magnetometer models. This is done using SystemVue's *Math Language* blocks.

The output of the sensor models are then elaborated by the microcontroller, executing the EKF algorithm. It produces an estimated orientation that can be compared with the original data provided by OpenSim, to evaluate the tracking error.

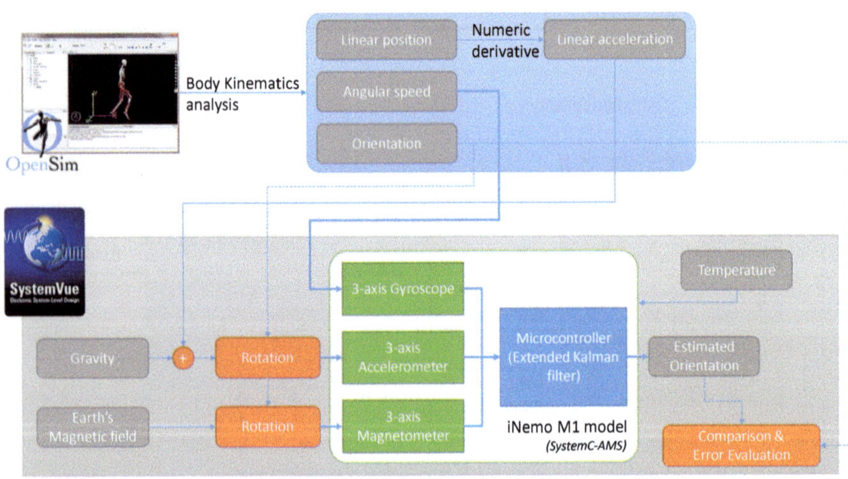


**Fig. 8.10** System-level simulation and analysis setup

## 8.3.3 *Experimental Results*

### 8.3.3.1 Sensor Model Validation and Calibration

The sensor SystemC-AMS models derive from the abstraction of lower-level, detailed models of the MEMS, analog and digital parts, as described in the other chapters of this book. In order to further validate the modeling methodology, characterization experiments were made to evaluate the accuracy of the model with respect to iNemo M1 prototype boards in static conditions. For all the experiments, the accelerometer is configured with ±2g full scale (g is 9.81 m/s$^2$, while 1 mg is $9.81 \times 10^{-3}$ m/s$^2$) and 50 Hz output data rate, the magnetometer has a ±1.3 G full scale and 75 Hz output data rate, and the gyroscope is configured with 250 dps (degrees per second) full scale and 95 Hz output data rate.

For comparing the performance of two sensors, or a sensor and its model, as in this case, calibration is required. In the case of a sensor characterized by a linear transfer function, two parameters have to be determined, i.e., offset and gain. One possible way is to apply known reference stimuli and then error minimization techniques (such as least square) to find the best line fit. However, when inertial sensors are concerned, it may be complicated and expensive to supply precise stimuli in a laboratory setup. As an example, the accelerometer may be calibrated using the Earth's magnetic field, and then the sensor may be aligned in at least two positions (e.g., horizontal and vertical) to measure 9.81 m/s$^2$ and then 0: the precise alignment can be hard to verify. In addition, when multi-axial sensor are used, the non-ideal alignment of sensing devices also has an impact on the resulting measure. Different calibration approaches for magnetometers and accelerometers have been developed when no precise stimuli can be applied [13]. Such methods are based on the fact that the measure of the magnetic field (or gravity) vector taken with a rotating tri-axial sensor should lie on a sphere centered at the origin; however, due to the sensor non-ideality, a translated ellipsoid is usually obtained. Geometric fitting algorithms such as the one in [14] can then be applied to find calibration parameters. Gyroscope calibration is more demanding, since the application of dynamic stimuli with known angular velocity can be an issue. For this reason, only the offset in static condition is evaluated.

Table 8.2 reports the sensor reading (average data on 60 s at 25 °C) before and after calibration for an available prototype, aligned on the horizontal plane and with the magnetic North, in comparison with the simulated standard models. It appears that the noise range is in the same order of magnitude, while the sensor reading is sometimes different due to different non-ideal transfer functions or to uncertainty in the applied stimulus. Calibration brings the readings of prototypes and models close to each other, within the variability implementation ranges that can also be found in the sensor datasheets.

**Table 8.2** Sensor calibration and model validation

| Accelerometer | Prototype | | Calibrated prototype | | Simulation | |
|---|---|---|---|---|---|---|
| Axis | Avg. [mg] | Var. [mg$^2$] | Avg. [mg] | Var. [mg$^2$] | Avg. [mg] | Var. [mg$^2$] |
| $X$ | −7.20 | 7.41 | −5.41 | 7.31 | 2.27 | 5.71 |
| $Y$ | 7.70 | 6.66 | −5.48 | 6.55 | −5.78 | 4.14 |
| $Z$ | 1175.90 | 19.19 | 995.08 | 18.42 | 994.39 | 4.82 |
| Norm | 1175.90 | 19.20 | 995.11 | 18.42 | 994.41 | 4.85 |
| Magnetometer | Prototype | | Calibrated prototype | | Simulation | |
| Axis | Avg. [mG] | Var. [mG$^2$] | Avg. [mG] | Var. [mG$^2$] | Avg. [mG] | Var. [mG$^2$] |
| $X$ | 258.03 | 3.53 | 249.84 | 4.15 | 225.48 | 7.41 |
| $Y$ | −169.75 | 3.86 | 2.61 | 3.74 | −0.08 | 5.10 |
| $Z$ | −298.69 | 4.44 | −359.38 | 4.70 | −410.57 | 7.61 |
| Norm | 394.72 | 4.85 | 469.49 | 4.42 | 468.42 | 7.35 |
| Gyroscope | Prototype | | Calibrated prototype | | Simulation | |
| Axis | Avg. [mdps] | Var. [mdps$^2$] | Avg. [mdps] | Var. [mdps$^2$] | Avg. [mdps] | Var. [mdps$^2$] |
| $X$ | −2104.51 | 10578.27 | 0.0 | 10578.27 | 5.76 | 11968.58 |
| $Y$ | −5092.85 | 16049.69 | 0.0 | 16049.69 | −22.74 | 11978.18 |
| $Z$ | −1895.66 | 16881.37 | 0.0 | 16881.37 | 5.18 | 10560.46 |
| Norm | −5829.76 | 16861.32 | 0.0 | 16861.32 | 24.02 | 5865.78 |

#### 8.3.3.2 Drift Analysis and Mitigation

One of the main problems affecting the performance of an IMU is drift. It causes a slow change of the estimated orientation that occurs also when the sensor node is in a fixed position. Its causes are due to the system non-ideality. As an example, if only a gyroscope is used for an orientation estimation system, one may propose to integrate the angular velocity given by the sensor to compute the angular displacement from an initial condition. However, the non-ideal sensor (e.g., whose reading is affected by a non-zero mean random noise), would generate a drift over time, whose magnitude is proportional to the noise mean value. The use of more than one sensor, in the so-called sensor fusion approach, can greatly help in the minimization of drift. As aforementioned, the EKF is an optimal estimator that allows minimizing the error components when multiple complementary data sources are available. The use of an accelerometer and a magnetometer in combination with a gyroscope provides data relative to a fixed reference system and helps reducing the drift effect in static condition (i.e., when the accelerometer only senses the effect of gravity) and with predictable magnetic field values. However, many parameters contribute to the final effect, also considering the specific application and mission environment, and the drift problem, even if with a reduced extent, may still exist. The specific conditions into which a sensor node is used and the environmental and positional constraints can be used to further improve the tracking performance. As an example, a sensor node constrained to a body limb has fewer degrees of freedom of movement and thus drifts on certain axes can be suitably filtered.

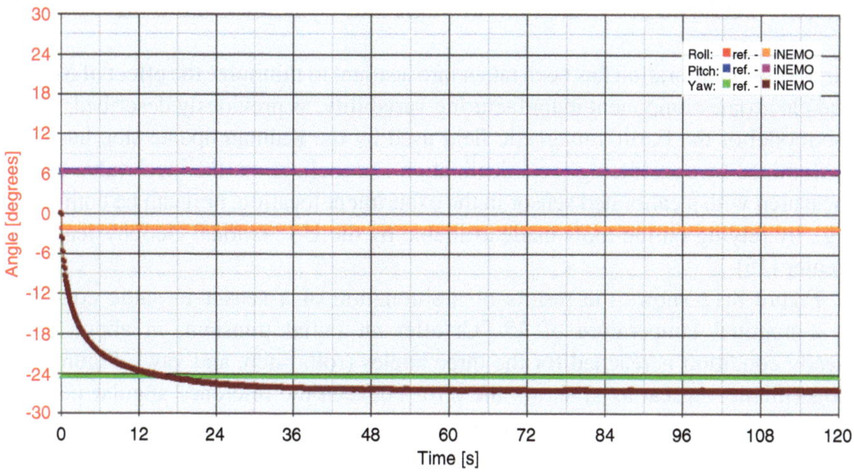


**Fig. 8.11** Estimation of Euler angles in static conditions, with an uncalibrated magnetometer: drift on the yaw axis is evident

The iNEMO M1 model enables studying the causes and then mitigating the effect of drift for a sensor node, where no geometrical constraints are considered. A large number of simulations were made to determine the most important causes of drift, taking into account sensor noise, imperfect calibration, environmental modeling (the Earth's magnetic field), and the internal parameters of the Kalman filter (noise covariance matrices). Figure 8.11 shows the estimation of Euler angles (roll, pitch, and yaw computed from the quaternions) generated by the virtual model running a 2-min simulation in static condition, where the magnetometer model is not calibrated and characterized by an offset comparable to the one of the real devices (about $\pm 50$ mG per axis). The combined effect of sensor noise and uncalibrated data feeding the Kalman filter has two main consequences: a particularly evident drift on the yaw axis, whose position estimation converges very slowly to a steady-state value (about 60 s are required), and a final yaw position estimation affected by a large error ($\sim 3°$). As an example, in a limb tracking system, the absolute error may be less relevant for the estimation of the relative position of each part of the body; the drift due to the slow convergence, instead, is an unwanted effect severely affecting the performance of body motion reconstruction.

Noise and uncalibrated sensors are among the main sources of drift, but the final system performance is affected by other causes as well, such as poor magnetic environment modeling, magnetic field disturbs and temperature changes. Taking into account such affects (as described and demonstrated in the following paragraphs) allows reducing drift to a negligible value for the targeted applications, i.e., reducing the convergence time of angular position estimation while maintaining a low absolute tracking position error.

#### 8.3.3.3 Performance Evaluation in Static and Dynamic Conditions

First, sensor calibration has been taken into account to minimize the effect of MEMS and electronic component manufacturing variability, as previously described. Then, the model of the Earth's magnetic field used by the Kalman update step has been customized by considering local declination and inclination: the local value can be measured with a calibrated sensor in the experiment location, or it can be computed, e.g., by relying on the tools made available by the US National Geophysical Data Center [15].

Figure 8.12 shows the results of a simulation of a sensor in static condition at a nominal temperature of 25 °C: after an initial transient, in about 3 s a steady condition is reached for the three angles (roll, pitch, and yaw). Figure 8.13 (logarithmic vertical axis) shows the error between the reference angular position and the values estimated by the emulated system, which, at the end of the simulation, is less than 0.5°.

Temperature has an impact on the estimated quantities as well. In the range between 0 and 50 °C, simulation shows that the Euler angle estimation error in static condition doubles with respect to the nominal temperature of 25 °C. Temperature compensation has been then introduced, by studying the simulated behavior of the sensors: for each of them, a linear correction function is defined, using the best fitting parameters derived from simulation. This keeps the error magnitude constant in the whole addressed temperature range, but requires an additional sensor in the system.

The developed simulation framework also enables evaluating the effects of the environmental magnetic field. Two effects are taken into account: the local field value, due to the position on Earth where the experiment is set, and the distortions introduced by local sources such as permanent magnets or electric currents in specific positions in space. In the first case, the local inclination and

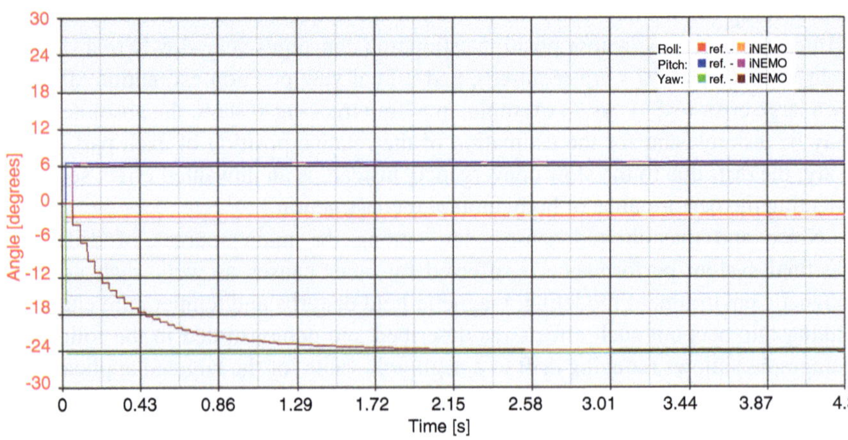


**Fig. 8.12** Euler angle estimation in static condition (input angles −2.37°, 7.5°, −24.13°)

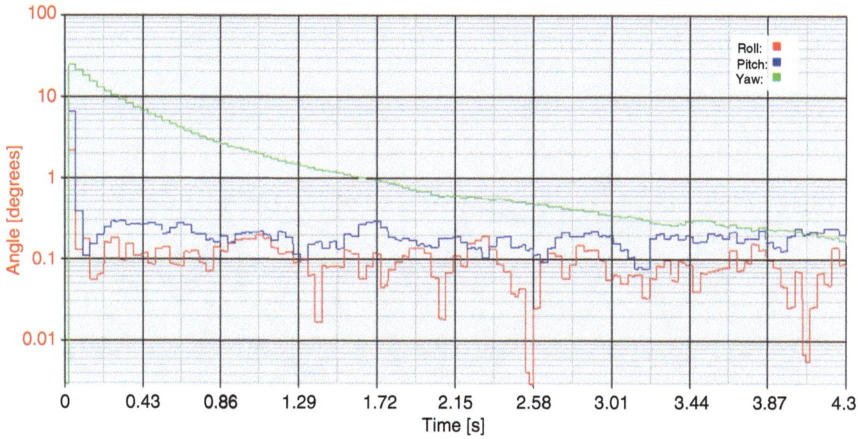


**Fig. 8.13** Error on Euler angle estimation (absolute value) in static condition

declination are defined in the SystemVue Math Language block computing the magnetometer stimuli; the difference between the local field and the model within the EKF can enlarge the absolute Euler angle estimation error and increase the filter convergence time. In the second case, local perturbations can be introduced by using a specifically defined SystemVue Math Language block, modeling the presence of a magnetic dipole characterized by its magnetic moment and position. To reduce the effects of local magnetic sources, it is possible to weight the magnetometer input by dynamically changing the observation covariance matrix, e.g., by increasing the parameters proportionally with the difference between norm of the magnetometer value and the expected one.

Finally, the parameters of the Kalman noise covariance matrices have been fine-tuned to further reduce the estimation errors. An optimal result is shown in Figs. 8.14 and 8.15, where a virtual sensor node is placed on the femur of the OpenSim human body model during a walking simulation. It appears that the model estimation of the angular values closely follows the reference values, with an average error that, after a short initial transitory phase, reaches under 1° on the three axes with no significant drift.

In conclusion, new high-level modeling and simulation approach, supported in this case by the SMAC Platform, is useful to evaluate the system performance. Differently from traditional approaches requiring complex and expensive laboratory setup, with the proposed approach, the software development and algorithm optimization can be handled in a simulated virtual environment, supporting the optimization of performance metrics (errors and drift) and the design and evaluation of new applications, in a previously unavailable manner.

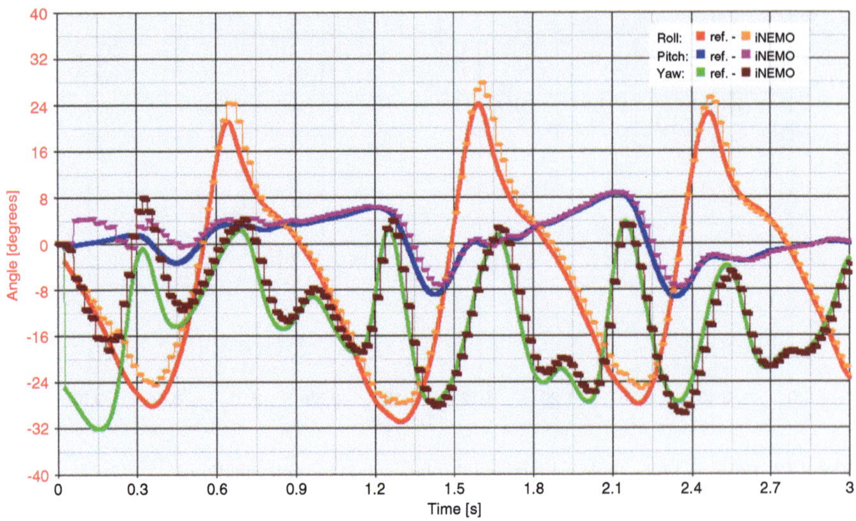


**Fig. 8.14** Euler angle estimation in dynamic condition

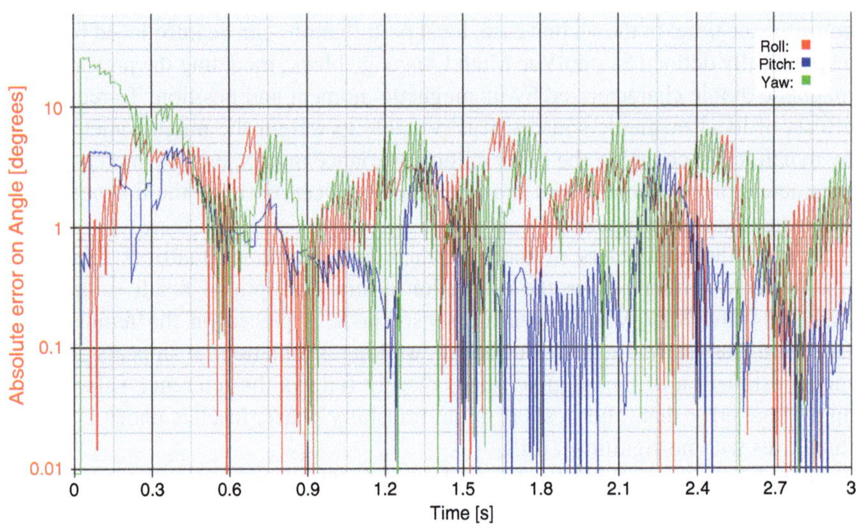


**Fig. 8.15** Error on Euler angle estimation (absolute value) in dynamic condition

#### 8.3.3.4 Model Performance

The data files including linear and angular position, velocity and acceleration of each selected "virtual" sensor position are generated with OpenSim at a frequency of 1 kHz. The final simulation model in SystemVue elaborates data at the frequency 1 kHz, which is much higher than the average human movement band (but required to correctly stimulate the sensor models), and feeds inputs to the sensors at this frequency. The SystemC-AMS internal engine of each sensor model runs at its specific frequency (up to 945 kHz for the gyroscope). Finally, the sensor fusion elaboration can operate at user-selectable frequency, here set at 50 Hz.

Each second of simulation of a single sensor node takes about 30 s of real time, on a quad-core Intel Core i7 – 2670QM running at 2.20 GHz with 4 GB of RAM. It has to be noted that each SystemC module instantiated within SystemVue runs on a separate thread, thus making the system scalable when more than one sensor node needs to be simulated, directly exploiting multi-core parallelism.

## 8.4 The Open-Source Test Case

### *8.4.1 Overview*

The OSTC aims at representing an example of smart system, providing sensing, computation, and communication. Thus, it is composed by a set of components belonging to different design domains. Figure 8.16 depicts its structure, while the different colors highlight the many design domains involved: digital HW (blue), analog HW (red), network peripherals (gray), system-level interconnections (orange), and embedded SW (yellow).

The components composing the platform are the following:

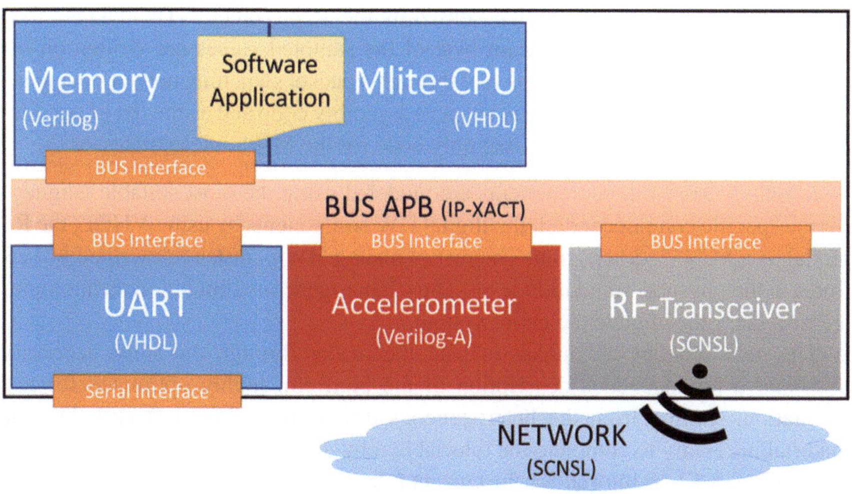


**Fig. 8.16** Overview of the OSTC, highlighting the different languages used

- *MLite CPU*: Digital HW component implementing a microprocessor compatible with the MIPS Instruction Set. This IP is originally provided as VHDL description.
- *Memory*: Digital HW component implementing a memory, it is used to store the SW application as well as the data sensed and computed by the platform. It also handles the communication with the peripherals using Memory Mapped Input Output techniques. It is implemented in Verilog.
- *UART*: Digital HW peripheral performing Parallel to Serial conversion. The IP is implemented in VHDL.
- *Accelerometer*: Analog HW peripheral. It is used to sense the environment, in particular the accelerations the platform is subjected to. The original description of the IP is provided in Verilog-A.
- *RF-Transceiver and Network*: Network peripheral used to transmit data over a packet-based network. It aims at modeling a wireless component, and it hence belongs to both the Network and RF design domain. The original model is developed using the SystemC Network Simulation Library (SCNSL) [16].
- *APB Bus*: Main bus used to connect the CPU-memory sub-system to the peripherals of the system. The bus used is respecting the ARM APB specification. The model is given both in Verilog and as IP-Xact specification.
- *Communication, Interfaces, and Interconnections*: The internal connections and the communication among the components of the platform are described using IP-XACT. This specification belongs to the electrical system-level design domain.
- A *Software Application* is running on the platform. The software is performing a reset phase taking care of configuring the UART module. Then, in its main loop it samples and gathers data sensed by the accelerometer. The data are then transmitted: the least significant byte of the sampled values are written on the UART, while the entire value is sent through a packet switching network using the RF-transceiver. The software is given as assembly code and as the list of opcodes obtained after compiling the assembly code for the MIPS microprocessor.

Initially, most of the components are modeled using the most suitable language for the specific component domain: digital HW components by using VHDL, the RF model as a SCNSL description. The rest of analog HW is instead described in VerilogA at the physical-simulation level. Thus, heterogeneous simulation technologies are necessary to simulate the entire system.

The next sections are meant to lead the reader through the steps necessary to obtain a homogeneous description of the platform, highly optimized for fast simulation, starting from the heterogeneous description of the test case. All the material necessary to complete the tutorial hereby described, is available at the web page in Fig. 8.17, as tutorial for the current release of HIFSuite.

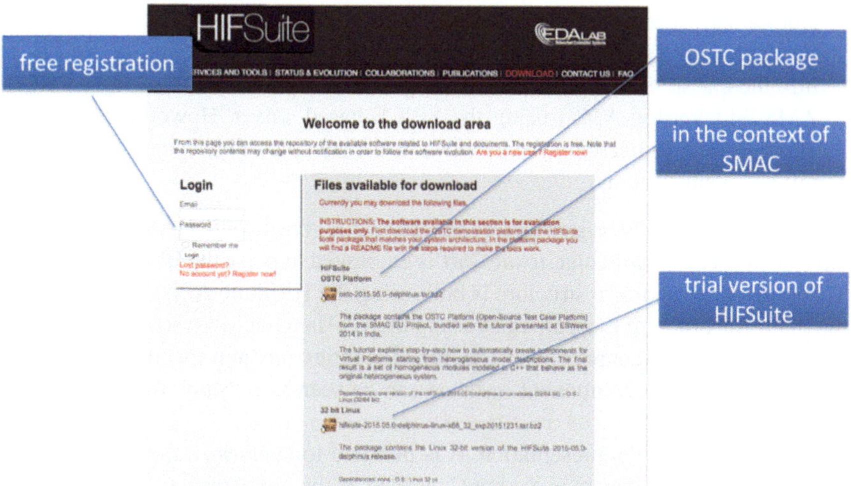


**Fig. 8.17** Web page containing all the material for the OSTC tutorial. Available at http://www.hifsuite.com/index.php/download

## 8.4.2 *Simulation of the Heterogeneous Description*

In order to simulate the heterogeneous description of the entire system, a discrete-event simulator is needed for the Digital HW, a SPICE-like solver for electrical equations is necessary to simulate the Analog HW, and finally a network simulator is needed for packet-based network simulation. Mentor Graphics' *Questa Advanced Simulator* [17] allows event-based simulation for classic HDLs (i.e., VHDL and Verilog) and SystemC using the *ModelSim* simulation technology. Mixed-signal simulation for the Analog HW is carried out by using the *ADMS* simulator that exploits a SPICE-based solver called *ELDO*. Network simulation is performed by SCNSL, an extension to SystemC to allow modeling packet-based networks.

However, some limitations arise in the simulation of the heterogeneous description of the whole platform by Questa Advanced Simulator:

- The system-level descriptions provided as IP-XACT models provide all the information needed to specify the system interconnections. However, the tool does not provide the capability of integrating the system using them. Thus, it is necessary to translate these interconnections into an accepted language.
- Network simulation is not supported. Even if the SCNSL library is based on C++/SystemC, some limitations imposed on the language by the tool do not allow to import SCNSL models. Some methodologies to co-simulate network and HW descriptions have been proposed in the literature, such as interfaces for connecting NS2 and ModelSim [18]. However, such techniques will require computationally expensive and complex SW interfaces based on TCP/IP sockets.

- Questa Advanced Simulator allows to mix classic HDLs (i.e., VHDL and Verilog) with SystemC within a unique model using ModelSim, as well as to mix the classic HDLs with their Analog-Mixed Signal extensions (i.e., VHDL-AMS and Verilog-AMS) using the SPICE-based solver. However, it does not allow to mix digital sub-systems described (even partially) using SystemC and mixed-signal systems.

In order to specify system interconnections, a set of tools based on HIFSuite have been used. In particular, *ipxact2hif* is employed to parse the IP-XACT models. Then, the entire platform structure is built using the HIFSuite APIs. After this step, the platform structure is specified by using the HIF language as a set of components interfaces and interconnections. The functionalities are not specified within the model, where every component instance can be seen as a "stub" for the actual IP implementation.

Starting from this representation, it is possible to synthesize the missing interfaces between components in SystemC, as well as the interconnection among them, by using the HIFSuite back-end tool *hif2sc*. The generated SystemC description is then the top-level of the entire OSTC. Since SystemC is an accepted language from Mentor's simulator, it can be compiled and linked together with the other digital components specified using VHDL and Verilog.

After this step, the digital part of the OSTC is ready to be simulated. The IP-XACT system-level descriptions can be imported into the simulation environment by exploiting HIFSuite capability of generating SystemC models. However, due to the limitations introduced before, a sub-system (even partially) specified using SystemC cannot be directly simulated with analog descriptions implemented using Verilog-A(MS). This is due to the fact that SystemC is supported only by the ModelSim simulation engine, while the mixed-signal descriptions require ADMS.

To overcome this problem it is necessary to instantiate both simulation engines and make them communicate. This can be done by splitting the entire system into two sub-systems. The first one is composed by the digital components, while the other is made of the analog parts of the system.

Once the system has been split into two parts it is possible to simulate both and make them communicate through a SPICE specification. In particular, a SPICE script is required to instantiate the top-levels of the two sub-systems and bind them by using electrical signals. Moreover, the SPICE specification must indicate the parameters required by the numerical solver for the analog part, such as the precision and the algorithm to employ.

Finally, it is possible to execute Questa Advanced Simulator without specifying any system, but using the SPICE description as command file. In such way, the simulator will take care of instantiating both available simulation technologies (i.e., ModelSim and ADMS), executing and synchronizing them by using a common notion of time.

Figure 8.18 depicts the execution of the system within Questa Advanced Simulator. In the left-hand side column of the main window it is possible to see the two different instantiated sub-systems: *yenvironment* and *ytop*. The former is implementing the mixed-signal part of the system and is simulated using ADMS.

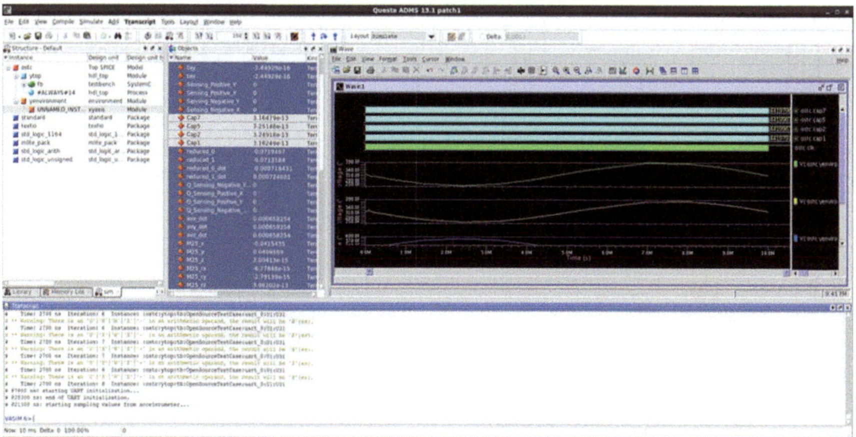

**Fig. 8.18** Execution of the heterogeneous system within Questa Advanced Simulator

The latter is the top-level module of the digital part of the system, simulated using classic event-based simulation. It is worth noticing that both are instantiated within the *OSTC* SPICE module.

A minor problem arises also from the use of SystemC as top-level of the digital sub-system. In fact, the simulation environment does not allow to specify a SystemC top-level in the SPICE specification. However, this problem can be easily overcome by wrapping the entire digital sub-system within a Verilog or VHDL component.

### *8.4.3 Automatic Integration of the Heterogeneous Description Through IP-XACT*

The first step in order to obtain a homogeneous SystemC description of the OSTC is to integrate the heterogeneous descriptions of its components. This is achieved by generating a top-level description from the automatically generated IP-XACT interfaces of the heterogeneous components, as shown in Fig. 8.19.

The descriptions of the components are translated to HIF descriptions by HIFSuite front-end tools. Then, the IP-XACT interfaces of the components are generated by the IP-XACT back-end tool. These interfaces are combined together in Kactus2 [19], an IP-XACT visual editor which allows to specify how the components are interconnected between them, thus creating the IP-XACT model of the whole platform. Figure 8.20 shows the IP-Xact model of the OSTC within the Kactus2 editor.

This IP-XACT description is then given as input to the IP-XACT front-end tool to produce the HIF description of the top level, which is then translated into a SystemC-RTL description featuring the instantiation of the components and their corresponding port bindings.

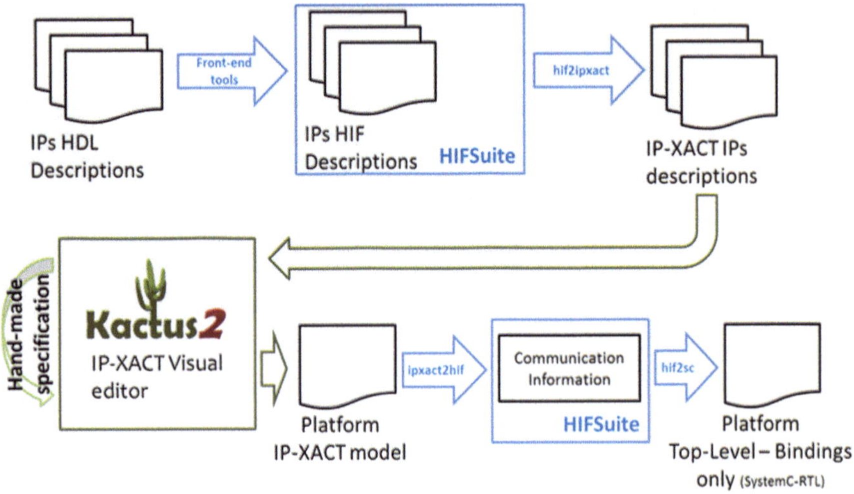


**Fig. 8.19** Automatic integration of heterogeneous descriptions through IP-XACT

## *8.4.4 Generation of a Homogeneous Description*

The next step consists of converting the heterogeneous descriptions of the components into a homogeneous SystemC RTL/AMS description. This is achieved by using HIFSuite front-end and back-end tools, as illustrated in Fig. 8.19.

The heterogeneous descriptions of the components are converted into HIF descriptions by HIFSuite front-end tools, and then converted into corresponding SystemC descriptions by HIFSuite back-end tools. Digital components are translated to SystemC RTL descriptions, while analog components are converted to SystemC-AMS descriptions. The final result of this step is a homogeneous SystemC RTL/AMS platform. Simulation of such a platform is performed by SystemC simulation kernels.

## *8.4.5 Automatic Abstraction and Data Type Optimization*

The homogeneous SystemC description of the OSTC can be optimized in order to improve simulation performance. Slow and inefficient SystemC data types are automatically replaced with native C++ data types by the HIFSuite tool *ddt*. *ddt* operates on the functionality code in the descriptions of the digital components of the platform. SystemC data types are replaced with corresponding native C++ data types. Data operations typical of HDLs are replaced with corresponding bitwise operations on the native C++ data types. This transformation improves simulation performance, at the expense of multi-value logic accuracy. This means that the

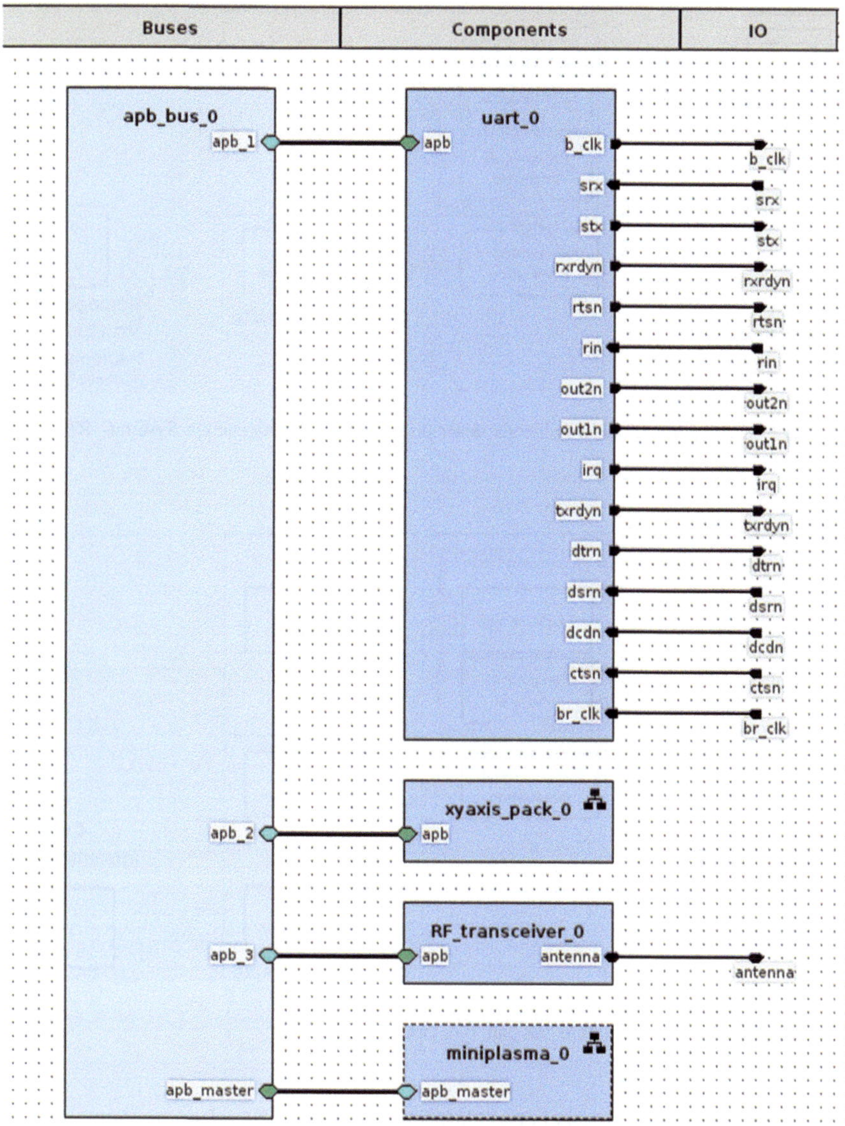


**Fig. 8.20** Automatic integration of heterogeneous descriptions through IP-XACT

4-value logic offered by logic SystemC data types is lost, since native C++ data types provide only 2-value logic. However, this turns out to be an acceptable loss of accuracy, since multi-value logic mostly affects only the initial state of the platform. Nevertheless, the presence of a reset phase in the software executed by the MIPS CPU ensures that simulation starts from a given known state, thus removing the need of modeling unknown states at the beginning.

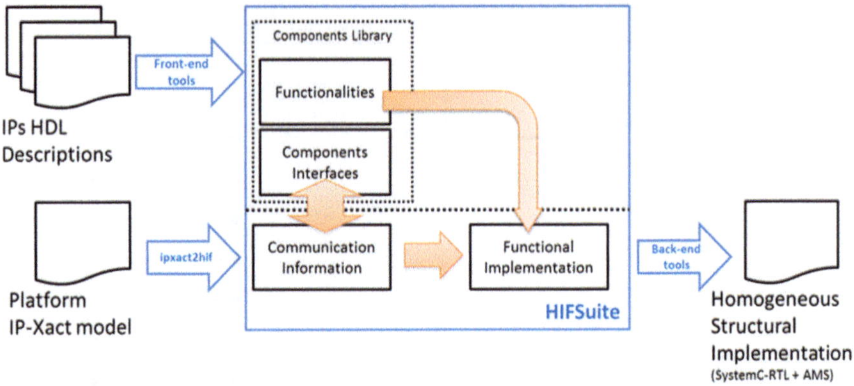


**Fig. 8.21** Conversion of heterogeneous descriptions into a homogeneous SystemC RTL/AMS description

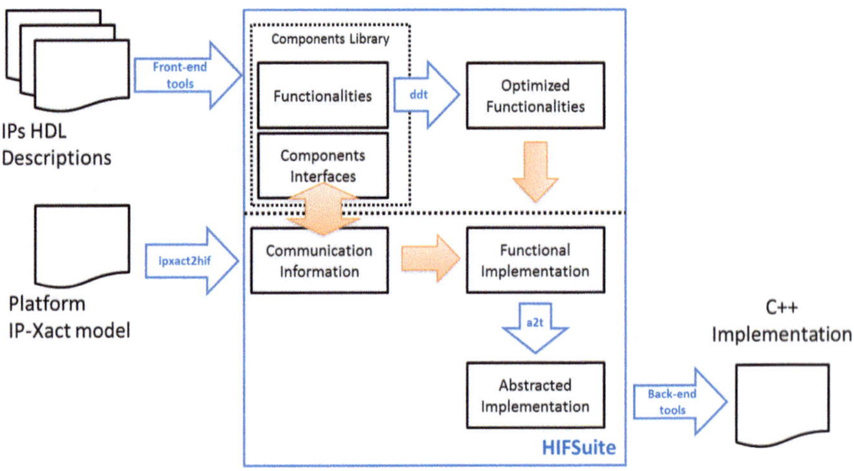


**Fig. 8.22** Automatic abstraction and data type optimization

Additionally, the homogeneous descriptions of the digital components can be automatically abstracted to corresponding transactional-level C++ descriptions by the HIFSuite tool $A^2T$. In this case, the SystemC simulation kernel is replaced by the lightweight C++ process scheduler automatically generated by $A^2T$.

Figures 8.21 and 8.22 illustrate the conversion, automatic abstraction, and data type optimization flows.

**Table 8.3** Simulation time using different simulation environments

| Version | Simulation time (s) | Speed-up |
|---|---|---|
| Heterogeneous Simulation (no network) | 215.47 | NA |
| Homogeneous RTL Simulation (no network) | 97.59 | 2.21× |
| Abstracted Model (no network) | 20.60 | 10.46× |
| Abstracted Model (with network) | 21.26 | 10.13× |

### *8.4.6 Simulation Results*

The simulation of the system using Mentor's simulation environment has been compared with the SystemC and C++ models of the OSTC obtained by employing the methodologies developed. The heterogeneous simulation has been performed with different SPICE configurations, while the SystemC and C++ homogeneous descriptions have been executed by using a sample rate fine enough to provide a high level of accuracy for the analog part.

Table 8.3 shows the execution times needed to perform a simulation of 100 ms on the OSTC using the different available simulation technologies. The *Heterogeneous Simulation* has been performed by using Mentor's Questa Advanced Simulator. *Homogeneous RTL Simulation* and *Abstracted Model simulation* have been performed by applying the steps previously described in this document. From the point of view of the abstraction level, the *Homogeneous RTL Simulation* is a completely equivalent simulation with respect to the *Heterogeneous Simulation* (i.e., no abstraction is involved).

Since the effort required prevents to simulate the network together with the system using Questa Advanced Simulator, also the Homogeneous and Abstracted simulations have been performed without considering the network. However, the abstracted model has been simulated also plugging the network model. It is worth noticing that the overhead due to the SCNSL-based network simulation is negligible.

## 8.5 Conclusion

The presented case studies show that smart system design can take advantage from a common workflow for the design and simulation, where different component-level design and implementation methodologies converge towards system-level homogeneous modeling and simulation. This approach, implemented in the SMAC Platform, enables finding solutions for real system-level problems that require the analysis of complex interactions between heterogeneous hardware and software. The use of the platform largely accelerates design space exploration, design validation, and performance evaluation with respect to what is achieved with traditional techniques. Smart system integrators have now the possibility of working with realistic yet relatively fast models of components and system deriving from detailed representations, thus avoiding (or reducing) possibly expensive iterations of prototyping and experimental characterization.

The developed methodologies and tools support the analysis of multiphysical, functional behavior of components and systems in most of the fields on which today's smart systems are based. The ground-breaking innovations pave the way for future development and have started the task of changing the way smart embedded systems are conceived and developed by design teams in an industrial ecosystem.

## References

1. R. Gillon, G. Gangemi, M. Grosso, F. Fummi, M. Poncino, Multi-domain simulation as a foundation for the engineering of smart systems: challenges and the SMAC vision, in *IEEE International Conference on Electronics, Circuits and Systems* (2014), pp. 858–861
2. M. Crepaldi, M. Grosso, A. Sassone, S. Gallinaro, S. Rinaudo, M. Poncino, E. Macii, D. Demarchi, A top-down constraint-driven methodology for smart system design, in *IEEE Circuits and Systems Magazine, First Quarter* (2014), pp. 37–57
3. M. Crepaldi, A. Sanginario, P. Motto Ros, M. Grosso, A. Sassone, M. Poncino, E. Macii, S. Rinaudo, G. Gangemi, D. Demarchi, Towards multi-domain and multi-physical electronic design, in *IEEE Circuits and Systems Magazine, Third Quarter* (2015), pp. 18–43
4. Pico Projector Market by Technology, Type, Product Model, Brightness, Application, Geography, Forecasts & Analysis (2013–2020). Research and Markets, 2014
5. N. Goren, I. Luft, S. Sourani, Method and device for monitoring movement of mirrors in a MEMS device. U.S. Patent 2011/0109951 A1, May 12, 2011
6. Vicon Motion Systems Ltd. http://www.vicon.com/
7. M. Grosso, G. Gangemi, S. Rinaudo, F. Cenni, M. Crepaldi, A. Sanginario, D. Demarchi, Enabling smart system design with the SMAC Platform, in *IEEE Symposium on Design, Test, Integration and Packaging of MEMS/MOEMS* (2015)
8. F. Cenni, O. Guillaume, M. Diaz-Nava, T. Maehne, SystemC-AMS/MDVP-based modeling for the virtual prototyping of MEMS applications, in *IEEE Symposium on Design, Test, Integration and Packaging of MEMS/MOEMS* (2015)
9. STMicroelectronics. iNEMO engine software libraries, http://www.st.com/web/catalog/tools/FM147/CL1818/SC1528
10. A. Sabatini, Quaternion-based Extended Kalman Filter for determining orientation by inertial and magnetic sensing. IEEE Trans. Biomed. Eng. **53**(7), 1346–1356 (2006)
11. C.M.N. Brigante, N. Abbate, A. Basile, A.C. Faulisi, S. Sessa, Towards miniaturization of a MEMS-based wearable motion capture system. IEEE Trans. Ind. Electron. **58**(8), 3234–3241 (2001)
12. S.L. Delp, F.C. Anderson, A.S. Arnold, P. Loan, A. Habib, C.T. John, E. Guendelman, D.G. Thelen, OpenSim: open-source software to create and analyze dynamic simulations of movement. IEEE Trans. Biomed. Eng. **54**(11), 1940–1950 (2007)
13. X. Zhang, L. Gao, A novel auto-calibration method of the vector magnetometer, in *IEEE International Conference on Electronic Measurement & Instruments* (2009), pp. I 145–150
14. Q. Li, J.G. Griffiths, Least square ellipsoid specific fitting, in *IEEE Geometric Modeling and Processing* (2004), pp. 335–340
15. National Geophysical Data Center – National Oceanic and Atmospheric Administration, Magnetic Field Calculator. http://www.ngdc.noaa.gov/geomag-web/

16. F. Fummi, D. Quaglia, F. Stefanni, A SystemC-based framework for modeling and simulation of networked embedded systems, in *Proceedings of IEEE Forum Specification, Verification and Design Languages* (2008)
17. Mentor Graphics, ModelSim-advanced simulation and debugging, 2012
18. U. Hatnik, S. Altmann, Using ModelSim, Matlab/Simulink and NS for simulation of distributed systems, in *International Conference on Parallel Computing in Electrical Engineering* (2004)
19. A. Kamppi, L. Matilainen, J. Maatta, E. Salminen, T.D. Hamalainen, M. Hannikainen, Kactus2: environment for embedded product development using IP-XACT and MCAPI, in *Euromicro Conference on Digital System Design (DSD)* (2011)

# Index




N. Bombieri et al. (eds.), *Smart Systems Integration and Simulation*,
DOI 10.1007/978-3-319-27392-1

MIX
Papier aus verantwortungsvollen Quellen
Paper from responsible sources
FSC® C105338

If you have any concerns about our products,
you can contact us on
**ProductSafety@springernature.com**

In case Publisher is established outside the EU,
the EU authorized representative is:
**Springer Nature Customer Service Center GmbH**
**Europaplatz 3, 69115 Heidelberg, Germany**

Printed by Libri Plureos GmbH
in Hamburg, Germany